P9-CBM-648

BRAVE
GREEN
WORLD

Unipress Books

www.unipressbooks.com
First published 2021

Published by MIT Press by arrangement with UniPress Books Ltd.

Commissioning Editor: Kate Shanahan
Project Manager: Natalia Price-Cabrera
Design & Art Director: Paul Palmer-Edwards
Illustrator: Robert Brandt
Picture Researcher: Natalia Price-Cabrera

ISBN 978-0-262-04446-2

Library of Congress Control Number 2020933238

Printed in China

The MIT Press
Massachusetts Institute of Technology
Cambridge, Massachusetts 02142
http://mitpress.mit.edu

BRAVE **GREEN** WORLD

HOW SCIENCE CAN SAVE OUR PLANET

CHRIS FORMAN & CLAIRE ASHER

The MIT Press
Cambridge, Massachusetts

Contents

Introduction

The rings of a tree tell a story. A story about the life of the tree, and the environment in which it grew. But what is humanity's story? Will it be a blackened layer of fossilized smartphones in the footnotes of geology? Or will it be a story like the Daintree rainforest in Australia—one of the oldest surviving forest ecosystems in the world whose current inhabitants boast a direct lineage thought to be over 100 million years old? To envision a 100-million-year-long story for humanity, we must imagine a world where every generation returns the materials they use to the soil, air, and oceans in a way that enables future generations to use that material too. A world without waste or pollution. The transition to such a system—called a circular economy—depends greatly on science. In this book, we explore the science that could take us there.

Our current way of living is destined to change fundamentally in the next few decades. Exactly how that happens is a decision that will be made collectively by all of us. The materials we use to create our new world will depend on the technology at our disposal, which will be determined by the science we perform today and the politicians we allow to govern us. Balancing the long-term prospects of other people's great-grandchildren against our own short-term interests is not a trade-off that many people give much thought to. But natural systems suggest there is a way to provide technological luxury to us all, at the same time as guaranteeing a positive future for everyone's children.

Imagine if our future electronic devices could grow from the buildings around us like a tree bears fruit, and throwing away an old device was more like composting a discarded apple core. Smart materials and innovative molecular manufacturing inspired by natural organisms could trailblaze the way to a fairer, more sustainable society that boasts a hyper-efficient, innovation-driven circular economy. An economy equipped to address the multifaceted challenges of climate change, biodiversity loss, and inequality, while bringing countless benefits for all of humanity.

In the following pages, we set out our vision of a future where ten billion people can flourish on Earth indefinitely, without exhausting our planet's raw materials or harming the other eight million species on the planet. Such a circular economy could even drive our expansion to other planets—and that is a tale we want to hear!

Nature: the fabric for life

Our story begins at the planetary scale, where we look at Earth as a single system, basking in solar energy, and consider the consequences that physical laws have had for living organisms. We learn that nature has evolved some pretty sophisticated systems for capturing energy, passing it around, and squeezing as much work out of it as possible. To do this, biological organisms have developed an incredibly advanced toolkit for processing energy and matter.

In comparison to nature, humans do a poor job of maximizing our use of the gargantuan supply of solar energy that flows through the Earth. Our astonishing achievements—like the internet or exploration of space—rely instead on accumulated solar energy stored in fossil fuels. We have become accustomed to extravagant expenditure of stored-up energy as we burn through our supplies faster than they can be replaced. But in a 100-million-year-long story, oil becomes a renewable resource—provided we use it more slowly than it regenerates!

The natural world demonstrates that excellent fabrication does not require profligate energy use and this realization focusses our attention away from energy to look at materials. If we could take a leaf out of nature's copy book and fine-tune our manufacturing processes to maximize the utility of solar energy, we could give all our natural capital a chance to recover.

Materials technology

Energy handling is deeply integrated into the very molecules from which biological organisms are made. Learning to emulate this property would enable us to channel sunlight directly into manufacturing—sidestepping the need to produce electricity—and enable us to grow complex systems, one molecule at a time. Such a monumental vision demands a considerable expansion of our current scientific and technological capabilities.

3D printing, in which a digital design drives an additive fabrication process, is the first step toward emulating nature's ability to incorporate information into materials. How will additive manufacturing help us reach circularity? The power of computers and artificial intelligence will no doubt help. In addition, an entirely new toolkit is being designed to manufacture materials; synthetic biology promises smarter, bio-inspired materials that can interpret and store information. These could be the versatile building blocks we need to replace our current, non-regenerative systems.

Greater than the sum of the parts

As we explore these technologies and others, we begin to think about the effect of adding them together. We discover that AI and automation could give us the tools we need to explore a vast cathedral of materials in which a phenomenon, known as emergence, leads to powerful advantages. Taking the technological story to its natural conclusion leads us to a realm where computation itself leaps out of the computers and into the materials that will surround us. We envisage a future where all manufacturing processes—from making smartphones to building houses—can be based on a generalized architecture of computation using standard chemical building blocks. These building blocks can be easily assembled, broken down, and reassembled at almost no cost, and waste materials produced in the process can by siphoned off to fuel other activities, in the same way that plants, animals, fungi, and bacteria have evolved to exchange energy and materials within the wider network of an ecosystem.

A brave green world

To construct a circular economy, many moving parts need to come together. We need to think not in terms of products, but in terms of systems. We are already seeing the first green shoots of technical solutions, and these delicate beginnings must be nurtured if they are to lead us to the next human epoch, where sustainable resources are readily available for everyone, and where our energy, agriculture, manufacturing, and waste systems coexist in a robust, mutually-reinforcing global framework.

Everyone has a part to play in realizing this future. Users can choose to buy brands that have transparent, sustainable supply chains; investors can choose companies with ethical manufacturing processes or innovative waste-recycling schemes; business-owners can create a market for better materials by choosing sustainable suppliers; NGOs can support local initiatives to decentralize production and teach repair skills; funding bodies can create initiatives to develop new materials, new manufacturing processes, and new distribution models; policy-makers can channel funding toward the development of a circular economy.

The next chapter of the human story will depend on the collective effect of all these decisions. For humans to make individual choices that add up to a viable future for all life on Earth as we know it, we need a universal vision for what that future will look like. *Brave Green World* describes one possibility, with a roadmap of how to get there.

I Planet Energy

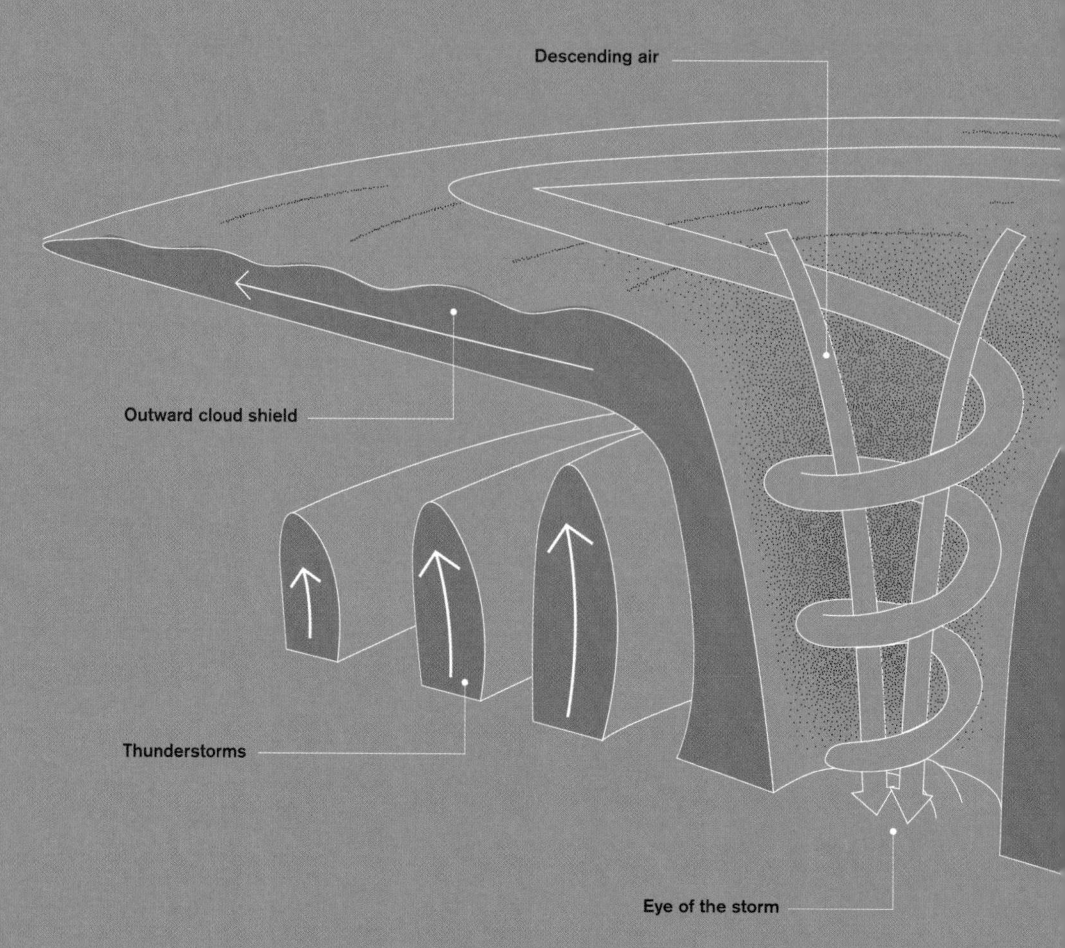

Hurricane structure

“The planet just glows. I remember trying to describe to my son, who was seven at the time, what it was looking like to me. I’m like, ‘Okay, the simplest way I can think is just, take a light bulb—the brightest light bulb that you could ever possibly imagine—and just paint it all the colors that you know Earth to be, and turn it on, and be blinded by it.’ Because day, night, sunrise, sunset, it is just glowing in all of those colors.”

Nicole Stott, Astronaut, NASA, March 2018

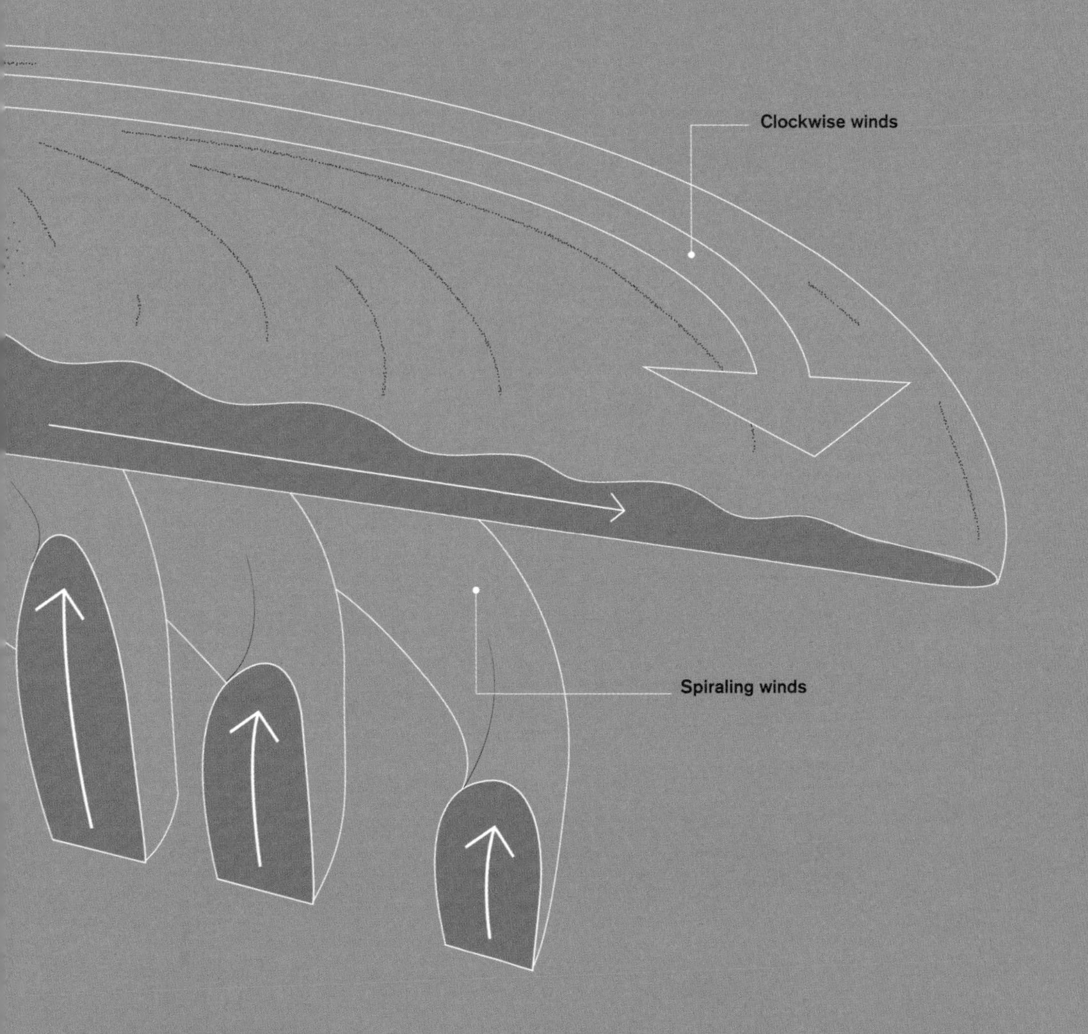

You may have heard of sunshine, but have you heard of "earthshine?" All day, every day, a vast and continuous flood of energy from the sun saturates Earth with light. Indeed, basking in solar splendor is a favorite pastime of almost every living creature on the planet. But while you're lying there sunbathing, have you ever wondered what happens to all that solar energy? The answer is surprisingly simple. It all leaves again.

Earth radiates nearly all incoming solar energy straight back into space as reflected light or heat—invisible infrared radiation—that we like to call "earthshine." About as much energy leaves Earth every second as strikes its surface, as if we were standing within a giant waterfall made from light. This luminous cascade of solar energy drives nearly all the natural surface processes that surround us—climate patterns, ocean currents, and the lives of organisms. All the colors in the room around you right now are produced by light on its journey back to space.

But if we are smart, we can store solar energy for a while —just as we can catch a handful of water from a waterfall and hold on to it for a moment before it slips through our fingers—and life on Earth has been doing this for around 3.5 billion years.

Where do humans get their energy?

Every single human endeavor is based on rearranging atoms using energy. In fact, there is practically nothing else that can be done in the universe *except* rearrange matter with energy. While some energy in the modern human world derives from nuclear and geothermal power, the vast majority comes, directly or indirectly, from sunlight. Farm crops convert solar energy into organic compounds via photosynthesis. Solar panels produce electricity and sunlight drives other renewable energy sources such as wind or the water cycle. And of course, prehistoric solar energy is stored in fossil fuels such as coal, oil, and natural gas.

By exploiting the colossal quantities of high-grade energy that has accumulated in these natural reserves over countless eons, and releasing it in the span of just two

Above: A view through the edge of the atmosphere from the International Space Station. We can see the air glowing from sunlight refracting through it, as well as a bolt of lightning reflecting off the space station.

centuries—a mere blink of the cosmic eye—human economic, agricultural, and industrial processes have spread with unprecedented speed across the planet. On a geological timescale, our massive global population expansion is virtually instantaneous, but already the immensity of human influence over natural systems rivals the magnitude of the biological world itself, and humans can alter the path of solar energy on its return journey to space through a variety of mechanisms. In truth, this is a phenomenal achievement—that also happens to be unsustainable.

Human activities modify natural levels of greenhouse gases in the atmosphere, principally water vapor (H_2O), carbon dioxide (CO_2), methane (CH_4), nitrous oxide (N_2O), and ozone (O_3). These gases naturally form a blanket around the Earth, which traps solar energy in the atmosphere that would otherwise be returned to space. Without this greenhouse effect, temperatures on Earth's surface would average around -2°F (-19°C) instead of a comparatively balmy 57°F (14°C). However, the flux of sunlight through our system is so huge that you don't need to change the concentration of greenhouse gases much to have a significant effect. The additional energy trapped in the atmosphere from human activity is sufficient to raise air and sea temperatures, cause ice sheets to melt, and sea levels to rise. It also fuels extreme weather conditions such as storms, wildfires, and tempestuous seas. For example, hurricanes are formed when warm oceans heat the air above them. The warm air rises, then cools, and sinks again. As it falls, the cool air is reheated by the warm ocean, forming a continuous cycle known as a convection system, which rotates due to Earth's own rotation and can create storms. The warmer the ocean, the more energy in the system and the more powerful the storm.

Opposite top: A star is a ball of plasma that is in a precarious balance between gravitational collapse and pressure generated by radiation released from nuclear fusion in the star's core. The charged matter in the sun generates complex magnetic fields that twist and writhe, often accelerating matter unpredictably spaceward in the form of solar flares.

Opposite bottom: If you vigorously swirl a glass of water and observe the resulting whirlpool, you can see that the more energy in a system, the greater potential for movement. The same is true for our atmosphere and oceans; powerful currents form between warmer and cooler regions resulting in giant storms like Gaemi in 2012, pictured here.

Physical boundaries

With a rapidly growing understanding of astrophysics, coupled with exploration of our own and other balls of rock that orbit the sun, we have become aware—with increasing clarity—how precious conditions on Earth are to life. Especially when we compare our pale blue dot with the inhospitable environments of other objects in the solar system.

A key set of indicators, termed the Planetary Boundary Conditions, has been developed to help us understand the atmospheric, chemical, and biogeophysical norms of our planet, so that we can track them and take actions to keep them within a habitable range. The level of CO_2 in the atmosphere, the extent of ocean acidification, and ozone depletion in the stratosphere are among these indicators.

We know that throughout geological history there have been huge changes in the Earth's systems. We are only just beginning to understand the magnitude of the influence that biological systems and humans can have on planetary conditions. For example, billions of years ago the development of photosynthesis released huge amounts of highly reactive oxygen into the atmosphere that other forms of life then had to adapt to.

Natural selection has tended to favor organisms that replenish the natural capital of their ecosystem. In the short term we have been able to exploit this natural capital—such as forests, fish reserves, soil fertility, oil, and so on—because the rocks, oceans, air, and soil have immense stores of resources. But we are exhausting that natural capital faster than it can regenerate, so it's just obvious that we will, sooner or later, be confronted by the full extent of our extravagant resource exploitation.

The international research organization Global Footprint Network calculated that in 2016 it would have required 20 months for Earth's resources to regenerate from a year's worth of human activities. This figure is predicted to rise to 24 months by the 2030s, meaning that humanity would require the equivalent of two planet's worth of resources to meet a year's worth of demand. There is only one Earth though. It's no different to balancing a check book. If you spend beyond your means, you will run out of money. Once the reserves are gone, the planet's natural regenerative capacity will not be able to support us anymore. Human activities are also encroaching on the natural world's ability to do its regenerative job, partly by displacing natural systems with human ones, but also by polluting the environment, placing it under further stress. Human operations

How much energy is there?

The total input of solar power into Earth is roughly 120,000 terawatts (TW), which is approximately 8,000 times more than the total global human demand for power. That's roughly the same number of solar photons per second as there are molecules in the atmosphere—enough power to run 80 trillion kettles continuously. If that energy didn't go back into space, things would get pretty hot around here pretty quickly! That's why only small amounts of greenhouse gases can make such a big difference to the temperature; they need to only trap a tiny proportion of solar energy to have a noticeable effect.

In fact, there's a reasonable chance that every exposed molecule on the sunny side of Earth is struck by at least one solar photon every 30 minutes. Each collision potentially produces anywhere between 1 and 1000 infrared photons driving potentially useful molecular rearrangements in the process. In chapter 2 we will explore in more depth how biological organisms can steer these rearrangements to their advantage.

The global human population is predicted to need 27 TW of power by 2050, growing to 43 TW by 2100, and the solar "waterfall" offers us up to 120,000 TW. At the time of writing, the best solar panels operate at about 47 percent efficiency—a number that changes frequently—meaning they convert less than half of the solar energy that hits them into useful electricity. The theoretical highest capture efficiency we can ever hope to achieve is 87 percent. Nevertheless, assuming that global demands for energy were to be supplied solely in the form of electricity—a discussion to which we shall return—then by 2050 we need only cover 0.05 percent of Earth's surface, or 230,000 square kilometers (89,000 square miles), to generate 27 TW. Based on these rough calculations, we can see that solar fields covering an area just one third the size of Texas could power all of humanity's activities well into the future. Texas supplies about 40 percent of the United States' production for crude oil—but with solar, this single state could *theoretically* supply the entire world's electricity.

Planetary boundary conditions

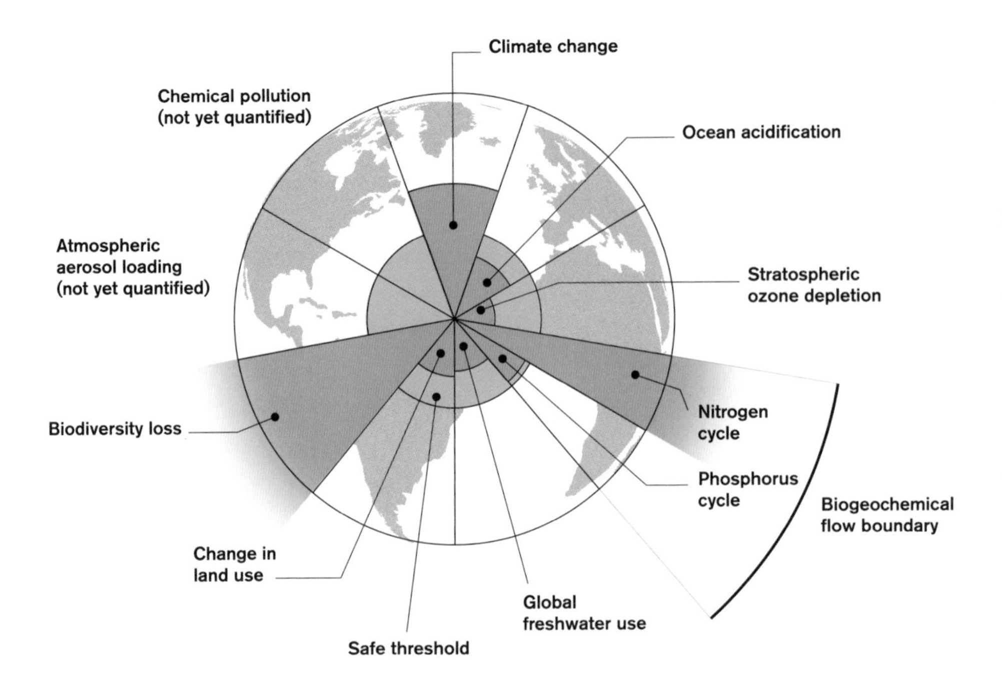

Above: Researchers have identified key indicators helping us to understand the effect we are having on the world around us. This framework helps governments set targets and goals that global industry must hit to ensure that our children can benefit from the current generation's legacy.

simultaneously reduce reserves and reduce the ability to regenerate reserves, so we need to change our operations to fit into the planetary boundaries as soon as we can. This book will demonstrate a range of technologies and sciences aimed at doing exactly that.

Combining wisdom, old and new

The philosophy that we should continuously regenerate our natural environment is neither novel nor extreme. Historically, there are many examples of people who fully understood and lived in harmony with the natural world for millennia, without exhausting the resources available or the capacity of ecosystems to regenerate. Fortunately, some modern communities have not lost this cultural legacy. As we develop new manufacturing techniques and materials, perhaps we can learn and take wisdom from all parts of the world to build societies that are nurturing and supportive, respecting of the environment and allowing humanity to progress economically, technologically, and scientifically. Perhaps even deepening our emotional and cultural connections with the natural world around us at the same time.

Cultures that understand sustainability, understand that organisms *are* the environment. Making changes is normal. Beavers construct dams. Birds build nests. Termites fashion vast, intricate, self-ventilating mounds. Herbivores shape trees and wolves can change the courses of rivers. But no creature is more adept at modifying its environment than an intelligent being.

Are we smart enough to live sustainably?

Intelligence has evolved independently in many different animal lineages, from apes to dolphins and octopuses to crows. We humans are clever enough to build rockets, power plants, and entire economies; smart enough to invent money, cars, and smartphones. But if the intellect of our own species ultimately results in accidentally tipping the planet beyond its boundaries for supporting life as we know it, then surely our intelligence will fail as a survival strategy? We are clever, but are we also wise enough to meet the challenge of powering our future civilization sustainably for 100 million years or more? What kind of technology or wisdom will help us achieve that? If that is our goal (and why wouldn't it be?), then we must each go well beyond thinking about only ourselves. Our choices today will determine whether we become a bitter but regretful species, a proud and arrogant one destined for self-destruction, or a kind, wise, and strong species, accepting our place in the world and facing the challenge of sustainability head on.

Whether we remain on Earth, travel to distant stars, or even colonize other worlds, solving sustainability is a requirement imposed by physics, not an optional extra.

Fortunately, humans have demonstrated a remarkable ability to grow in technological and economic sophistication. We need to better understand the way natural ecosystems operate so we can build our own ecosystems within them and manage natural systems effectively. That way we will be able to regenerate the natural capital on which the entire biological world relies, and in doing so, support the automatic regulation of our atmosphere and oceans. In such an economy—a circular economy—human systems and natural systems are mutually supportive, and all waste is recycled.

Our goal is to identify a kind of infrastructure that makes it easy for humans to make good, possibly selfish, short-term decisions while *automatically* operating in harmony with the geophysical and biological environment, leaving a positive legacy for our descendants. We look at the relevant scientific foundations and, step-by-step, explore current research that we believe will enable our species to develop wiser technology that will support and empower everyone to live harmoniously with their surroundings.

The fundamental laws of energy

If *all* the energy that reaches Earth's surface just radiates straight back into space, how can we use that energy? Whenever a photon of light interacts with a molecule, there is a chance that the molecule will absorb the energy of the photon. If that happens, then—for the briefest of moments—one of the molecule's electrons will become excited to a higher energy level, and from there a range of previously unavailable behaviors exist for the molecule. New modes of more energetic vibrations might become accessible, or parts of the molecule might be able to rotate. Each new possibility offers a different means of returning that energy to the universe. Such energy can never be destroyed—it has to go somewhere! Perhaps, the energy contained in the initial photon of visible light will be fractured into hundreds of infrared photons as the molecule vibrates, or stored chemically in a new molecular bond, or re-emitted as a single photon, albeit moving in a different direction to the original.

This movement of energy from hot bodies like the sun to cooler bodies like Earth, and then back out into the cold depths of space is known as thermodynamics—which literally means "heat movements." The principle that energy can never be created or destroyed is known as the first law of thermodynamics, and it applies to light as well as matter.

A single blue solar photon may contain enough energy to produce a hundred infrared photons during an interaction with a molecule. Therefore, the outgoing infrared photons from Earth considerably outnumber the incoming blue photons from the sun. This means that the number of possible ways to arrange the outgoing set of infrared photons—a characteristic known as entropy—is far higher than the number of ways to arrange the single incoming blue photon. A simple analogy is to imagine a stack of identical coins, each coin representing a unit of energy. Since all the coins are indistinguishable there is only one way to place all the coins in a single stack. But there are many ways to organize several smaller stacks that add up to the same value. And if we were to randomly assign an arrangement (perhaps by shaking the stack of coins in a container and tipping them onto a table), we would be far more likely to encounter a situation where we had a heap of unevenly distributed coins—possibly with several distinct stacks—than one perfect stack of all the coins arranged neatly.

Therefore, after a photon strikes a molecule and shatters into hundreds of infrared photons, it is very unlikely that those same infrared photons will ever again recombine as a single blue photon. The entropy of the universe has permanently increased. This tendency for entropy to increase is known as the second law of thermodynamics. Any kind of system will always tend to change in ways that increase the number of possible configurations available to it.

In the course of interacting with a molecule, the entropy of the light may have increased, but the entropy of the molecule does not necessarily change. If the interaction changes the molecule's state—say, by making or breaking a molecular bond—it might increase or decrease the entropy of the molecule. Just as a river can drive a water wheel round many, many times, a continuous stream of sunlight can cause the same molecule to rearrange repeatedly. We see that it is not just energy that we need to consider—it's the total change in entropy of both radiation and matter that tells us how many molecular changes we can make—think of it as our entropy budget.

From all this detail, a bigger picture emerges. By carefully controlling when and where molecules emit their infrared earthshine, organisms have evolved to exploit the constant flow of entropy into space to power their cells and find molecular configurations that are useful. For example, when sunlight strikes a plant leaf, photosynthesis traps that solar energy in a new molecular bond in a sugar molecule. That stored energy in the sugar molecular can sit patiently, waiting to be used to drive a molecular manufacturing process whenever the organism requires it. When that sugar is used, it gives out heat and also does some molecular "work" within the organism. Some of that energy will remain embodied in the organism's tissues, and some will go to space as infrared radiation, increasing the entropy of the universe, never to be seen again.

The amount of time that solar energy spends on Earth can vary enormously. A photon that hits a reflective surface and is re-radiated straight into space may spend barely a millisecond on our planet, but if a living organism is able to store that energy in the chemical bonds of a biomolecule, then the energy could remain on Earth, being passed from organism to organism, for thousands, even millions, of years—especially if it were incorporated in coal or oil.

Matter in motion

A molecule excited in the cascade of solar energy may strike its neighbor, passing on some or all of that additional rotational, vibrational, and translational energy. The result is that matter—even in its solid state—is filled with constantly jostling molecules and atoms that bash into one another causing random waves of energy to flow through every material. Rather like attending a rock concert, when hundreds of overly excited and energetic fans jump around and bump into one another, waves of compression can pass through the crowd. These waves are unpredictable; they just happen and they push you around. If you try to resist them, then you might cause the wave to change direction, but probably not.

In molecular terms, this random, excited motion is known as Brownian motion, named after the scientist who discovered it. Like waves on the ocean, some molecular waves are larger than others. Occasionally, a fluctuation passes through a material that has enough energy to cause a molecule, such as a protein, to change its shape. The compounds living organisms use—such as proteins, sugars, and DNA—are typically repeating molecular chains known as polymers that can alter shape in response to the Brownian motion of their surroundings. Such matter is known as soft matter.

Soft matter vs. hard matter

We know that certain living organisms can be wet, sloppy, and messy (athough we wouldn't describe trees in those terms!). In contrast, hard metals or concrete stay stubbornly in the same shape at room temperature. The reason these materials are so different has to do with an idea called an energy barrier.

Earthshine reveals material properties

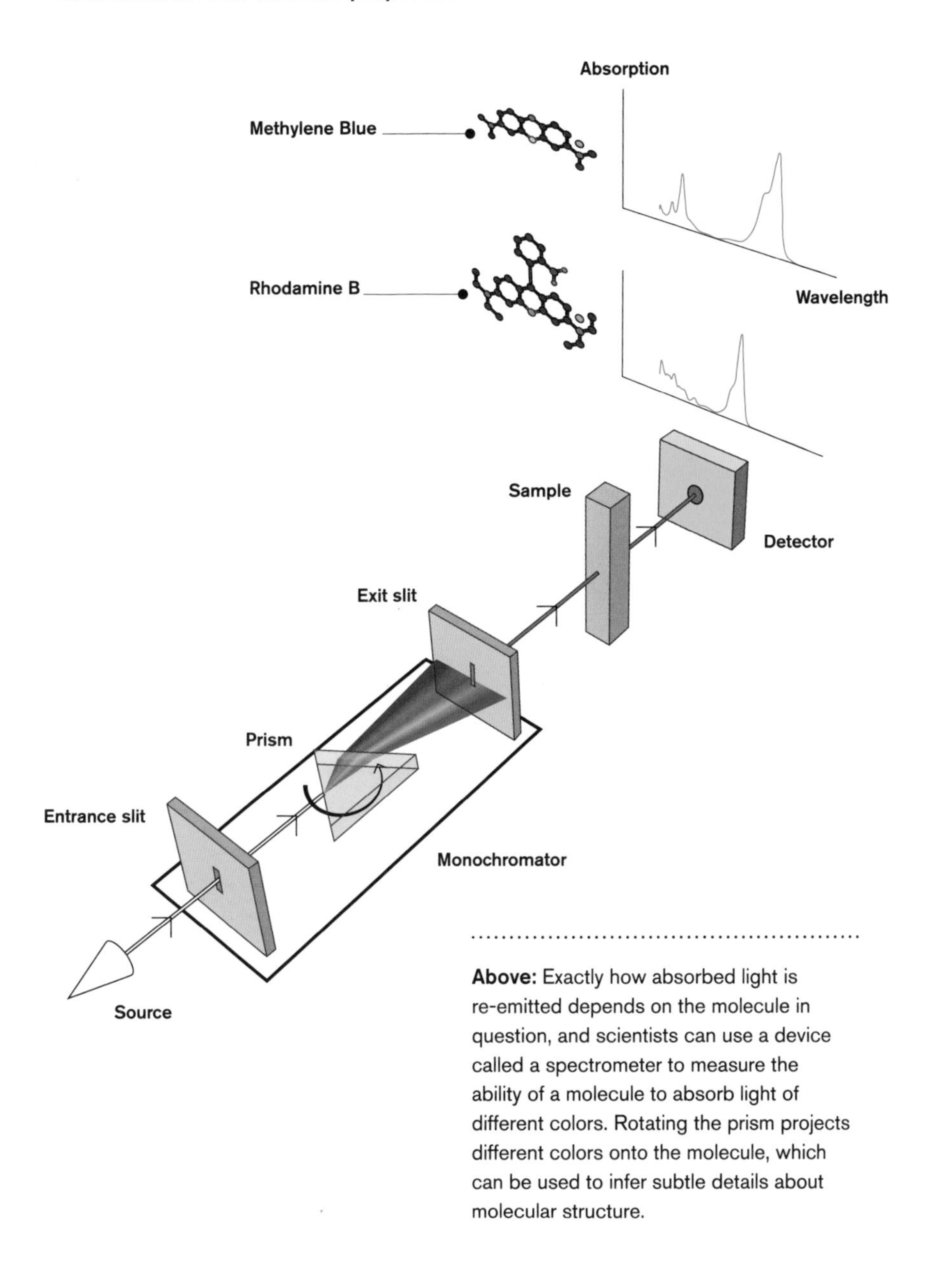

Above: Exactly how absorbed light is re-emitted depends on the molecule in question, and scientists can use a device called a spectrometer to measure the ability of a molecule to absorb light of different colors. Rotating the prism projects different colors onto the molecule, which can be used to infer subtle details about molecular structure.

Energy barriers represent the amount of work we must do to reorganize material. A simple example of an energy barrier comes from the children's toy, Lego®. If two blocks of Lego® are attached, the interlocking studs (on the top) and tubes (on the bottom) of each brick hold them together. We have to expend energy to prize them apart. Something similar happens for atoms and molecules in materials. Each atom is held in place by an invisible bond that keeps it in a particular configuration until a Brownian wave comes along to jostle the atoms into a new arrangement. For a metal, the energy barriers holding atoms in place can be very high, which is why it takes a lot of energy to make iron or aluminum sufficiently malleable to work with. In contrast, for biological materials, such as proteins or DNA, there are two levels of energy barriers. The barriers that hold the atoms together in molecules are very high; the bonds are very strong. But the barriers that control the shape of the molecule and how the molecules stick together are much lower. That is why low-intensity energy like sunlight or Brownian motion can manipulate biological matter, but high-intensity energy sources such as fossil fuels are needed to manipulate metals into useable forms.

In fact, the energy barriers that govern biomolecular rearrangements are tuned to the size of the waves of Brownian motion in biological material. That's one way of defining soft matter: materials whose energy barriers are about the same size as the energy fluctuations they normally experience. Even at room temperature, we don't need to input much energy to alter the shape of soft matter. If we could use soft matter in our technology, suddenly our manufacturing and recycling processes could become a lot less energy intensive, providing we can find materials that do the functions we require of them. Fabricating products from soft matter, in a manner more akin to biological processes, could allow us to use sunlight as an energy source for all our activities. Thus, the use of soft matter in manufacturing—already an emerging field—introduces entirely new avenues of material properties and fabrication processes.

Geothermal energy and nuclear power

Solar power isn't the only energy source we have at our disposal. Some of the energy we use derives from nuclear and geothermal energy sources, both of which—ultimately—come from the decay of matter into radiation. For example, the nuclear fusion occurring in the core of the sun and the nuclear fission taking place inside Earth's core both convert mass into energy, releasing photons. The result is that our planet is awash with geothermal energy from its core, as well as solar energy from above.

The total geothermal energy output of Earth's core is estimated to be 45 TW—almost double the need predicted for 2050—so theoretically we could source our power from the core, but in practical terms this would be extremely expensive in many locations.

Nuclear power requires less land area than any other form of energy production, but the process of nuclear fission produces lots of high-level radioactive waste that

Categories of soft matter

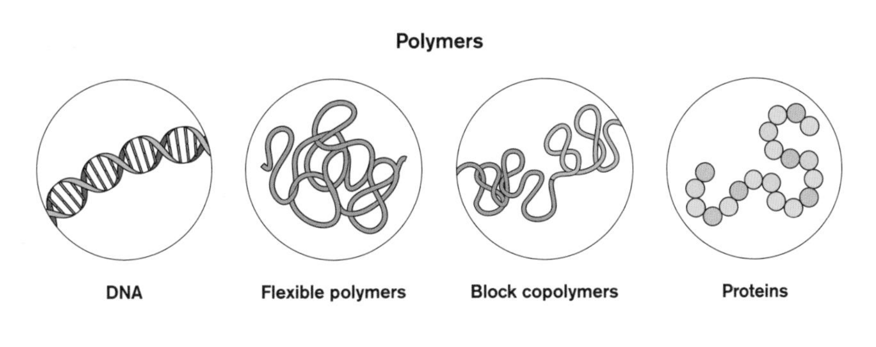

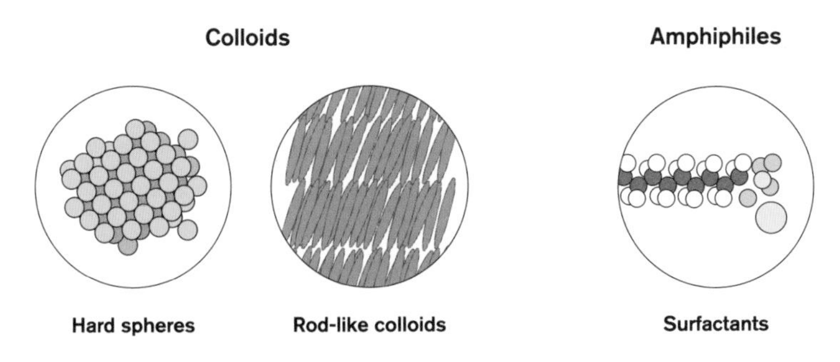

Above: Soft matter is easy to reconfigure at room temperature. There are several classes of soft matter that exist under Earth's conditions. For example, colloids are suspensions of particles in fluid, polymers are long flexible chains of repeating molecules, and amphiphiles are molecules that can sit at the interface of different fluids.

is extremely difficult—although not impossible—to dispose of. There are huge challenges associated with the safe disposal of radioactive fuel rods and contaminated waste materials.

Solutions can be found for relatively small quantities of nuclear waste, but scaling up nuclear energy generation to keep pace with humanity's insatiable demand would soon become unmanageable. To completely replace fossil fuels with nuclear power in order to achieve net zero carbon by 2050, there would need to be one nuclear reactor built every day, starting now. We could sidestep that problem and still meet most of our energy needs if we could make a success of nuclear fusion power, a process that is capable of providing abundant, clean, guilt-free energy. But so far that endeavor has proved elusive.

Perhaps that's for the best. Rather than propelling us to a sustainable utopia, in our current economic system such boundless quantities of fusion energy might accelerate our demise, by facilitating rapid expanse of consumerism at the expense of biodiversity and natural ecosystems; especially if it was labeled as "guilt-free" energy. Using a lower intensity energy source—such as sunlight—would force us to be more frugal with our energy use in the first place. The need to find material efficiencies would drive innovation in technology and manufacturing, forcing us to master the subtleties of manipulating energy barriers in materials; a pre-requisite skill for living sustainably on Earth or other planets.

Explosive energy usage

Regardless of the energy source, our current industries are simply not as efficient at choosing when and where to rearrange atoms as biological systems, which as we will see in the next chapter, have been fine-tuned through billions of years of natural selection. Our incredible technological revolution has been powered by high-grade, single-use fuels—mostly coal and oil. By using these fuels to generate electricity, we rapidly release large amounts of stored energy as earthshine in one go, using up our precious entropy budget. Often we use that energy to create relatively unsophisticated objects, such as aluminum ingots, which require further processing and manufacture.

It's becoming clear that what is important is not the total amount of energy, but the rate at which that energy can be applied to matter to reorganize its shape. The bottleneck here is the rate of relaxation and reorganization of useful molecules, which is determined by the energy barrier of the particular material we are using, and that is what dictates the

Geothermal energy

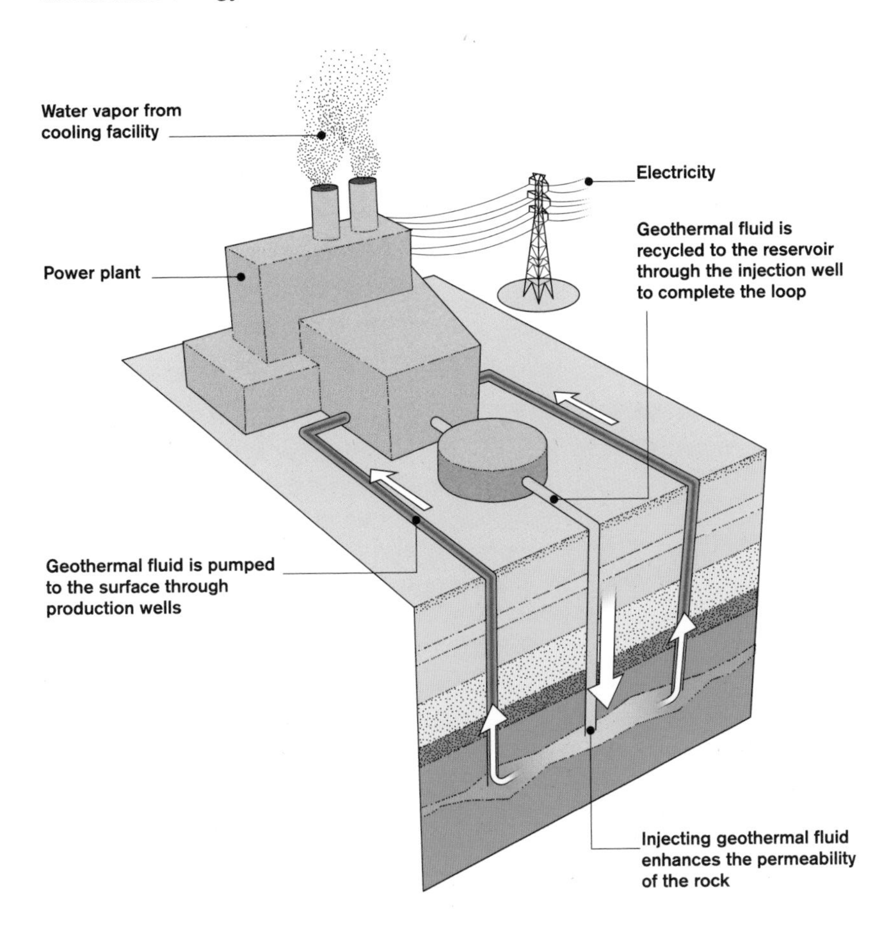

Above: Geothermal energy could in principle supply enough energy for humanity, but it becomes more expensive the deeper we delve, and requires specialized skills and equipment. The geology is not suitable everywhere to supply geothermal energy.

appropriate form of energy to use. If the rate of energy application is too high for the material, then it will burn or explode. If the rate is too low, no useful changes will occur. The energizing process must be compatible with the material's properties—which is why we heat steel in thousand-degree blast furnaces, but gently temper chocolate at around 86°F (30°C).

To highlight how clumsy our use of energy can be, compare the destructive power of a bomb with the amount of energy a human body uses per day. A healthy adult human may consume around 2,000 calories (cal) a day, which approximates to 8 megajoules (MJ)—about the same amount of energy as contained in eight sticks of dynamite. In a single day, a human expends eight times the amount of energy released when a single stick of dynamite explodes, but crucially the time taken to release that energy is vastly

Above: The interior of the Joint European Torus—a nuclear fusion reactor at Culham Center for Fusion Energy in the UK. Nuclear fusion promises cheap, abundant, side-effect free energy, but our attempts to develop a viable system have yet to succeed. Applying fusion power without discipline could destroy our ecosystems extremely quickly; applied wisely it could be used to recycle any material. The skill to choose between these outcomes could be learned by building a solar-powered economy first.

different. Instead of a few milliseconds, we take an entire day to convert 8 MJ of energy into a combination of earthshine and waste materials. Release that quantity of energy quickly and the results are explosive. A slow release allows living organisms time to put the energy to work, rearranging molecules in the body to produce helpful ones that can power metabolism, growth, and movement.

Clever manufacturing is all about pace—balancing the entropy production rate with the rate of energy release necessary to manipulate molecules into useful structures—and that timetable is governed by the properties of the material, not the energy source. It is therefore possible to imagine all our manufacturing processes on a scale of energy utilization, based on each technique's rate of entropy production and material reorganization costs, ranging between two extremes: the rapid, random release of energy from a lighted stick of dynamite and the slow, painstaking growth of an organism.

Right now, instead of making the best use of the entropy we export as radiation into space, we dump entropy into landfill in our own backyard, and as excess CO_2 and other emissions to the atmosphere. We thereby create a quadruple whammy—not only do we pollute the planet, we also fail to make the best possible use of our entropy production, increase the amount of energy locked in the atmosphere, and reduce the ability for natural systems to correct our errors. Perhaps then, our unsophisticated approach to materials is the greatest lost opportunity of our age for productivity and economic growth. We have a great deal to gain through improving material efficiency rather than blindly increasing the volume of limited resources we extract, consume inefficiently, and all too quickly discard.

Circular economics

Ultimately, a company's profit depends on how efficiently and intelligently they reorganize matter with the energy they have. However, if we continue to choose to reorganize matter in ways that cannot be sustained by the available supply of energy, our systems will collapse. Likewise, if we choose a rate of energy production that exceeds the rate at which we can successfully reorganize matter, we build up material entropy and drown in unusable waste. An alternative approach that would reduce the total impact of our activities on Earth is where the resources we extract and use are replenished and industrial waste products are passed on to a symbiotic industry for further use—the circular economy. If the world is unable to embrace this approach, we risk disrupting the very processes on our planet's surface which we rely on for our survival.

Futures: The smartphone case study

The biosmartphone thought experiment

We know that plants, through photosynthesis, intercept the flux of solar energy and use it to reorganize their molecules usefully into stems, leaves, flowers, and roots, before it is returned to space. All of which begs the question: what if we could learn to grow materials and products the way that wood, bone, shell, silk, and other materials are grown in the natural world? Could humans devise a system in which solar energy is channeled directly into manufacturing without the sidetrack of producing electricity first; where products are produced locally from recycled materials? How incredible would it be if a smartphone could be grown like an apple on a tree?

In this thought experiment, dispersed throughout the book, we take the example of a smartphone and look at how we might develop a biologically-inspired equivalent. We will analyze our hypothetical biosmartphone in the context of the main themes of each chapter, and by the end of the book we will have found examples in biology of nearly all the components and functionality found in a contemporary smartphone. As the book progresses, we postulate ways in which local waste materials and energy could be used to grow such a device in a manner reminiscent of the way that a tree grows fruit.

What components will we need to include in our biosmartphone? If you take a regular smartphone to pieces, you will find only a handful of key top-level systems: it needs energy (a battery), short- and long-term memory storage (RAM and flash storage), a user interface (camera, screen, buttons, speaker, microphone, vibration), a network interface (radio systems), a control system for linking components (processors and software), and an outer casing to hold it all together. From squid skin to silk, DNA to cellular biochemistry, we can take inspiration from nature to build each of these components.

We also need to be able to find a way to produce our device. In modern electronics all the components are manufactured by separate companies and integrated later. In a device grown in one place, all the components must be made and integrated at the same time—another major challenge!

Perhaps even more astonishing is the fact that nature can make all these components from basic but ubiquitous resources—air, soil, water, and sunlight—and at the end of their natural lifespan, they are automatically returned to these reservoirs, ready to begin the cycle anew. If we could harness and emulate these biological capabilities then we would be able to drive economic progress by powering manufacturing processes directly from sunlight, recycle all our products as easily as making compost, and re-task our manufacturing system to produce different products locally, as simply as changing DNA sequences.

We picked smartphones as an example in this thought experiment because they are among the most sophisticated products that we can make—so, if we could grow and recycle a smartphone in this manner, perhaps we could grow and recycle *anything* this way—maybe even starships.

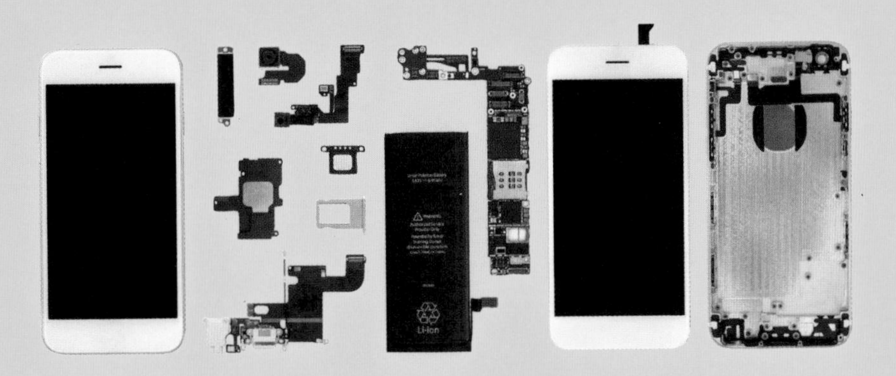

Above: Our biosmartphone would need to emulate the capabilities of a modern phone, including a battery and circuitry, RAM, data storage, a touch-sensitive screen, a speaker, a microphone, a camera, vibrating alerts, radio communications, and an outer casing.

2 Circular Ecology

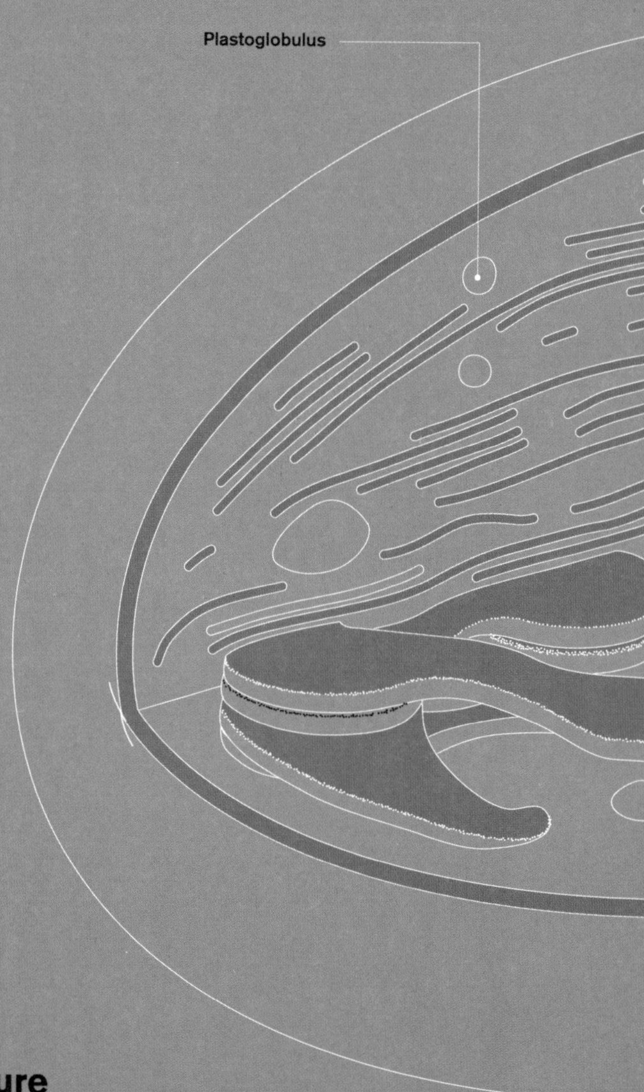

Chloroplast structure

“For the person for whom small things do not exist, the great is not great.”

José Ortega y Gasset, Spanish philosopher (1883–1955)

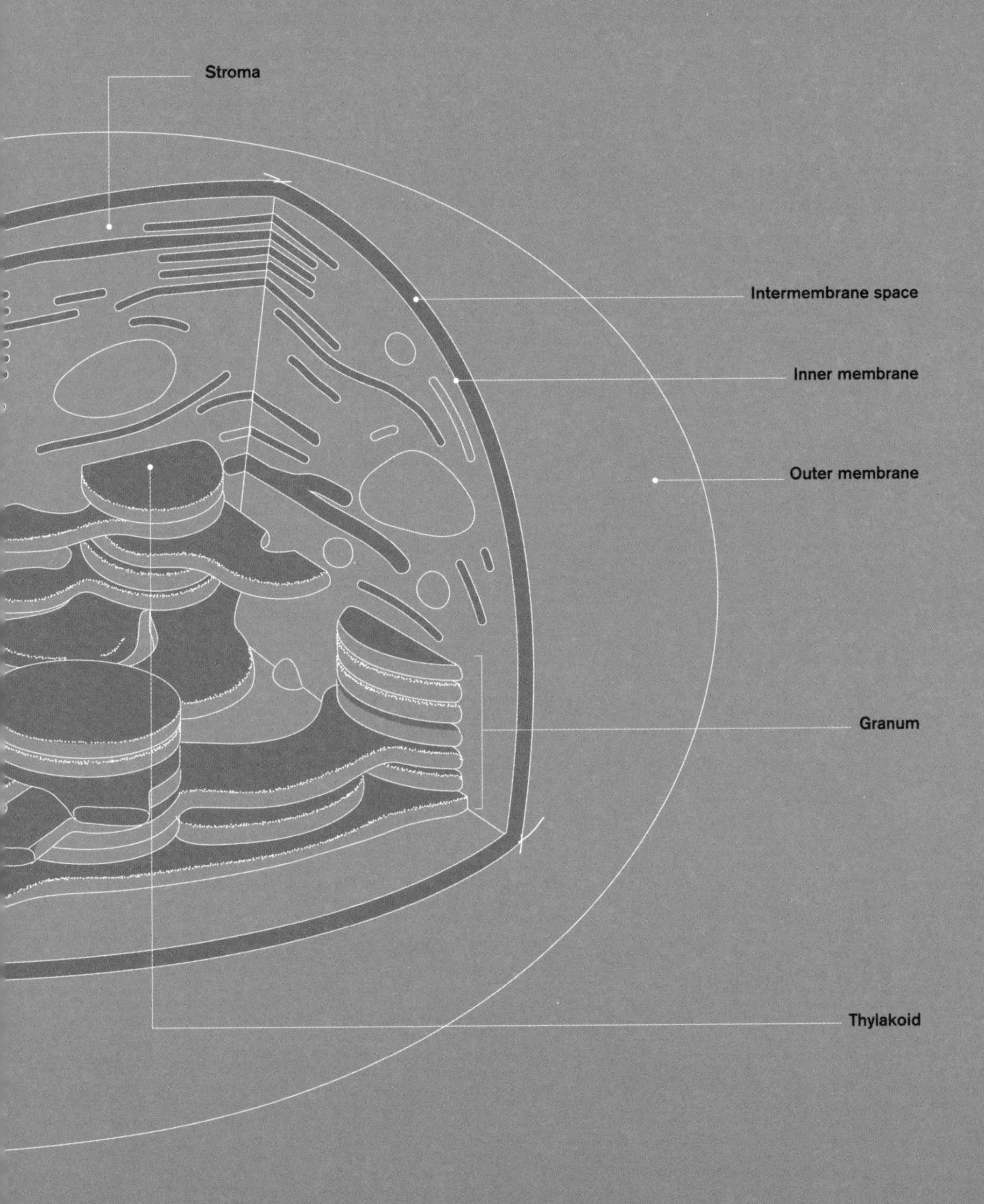

Nature offers a tremendous range of inspiration to help us imagine what a circular economy might look like. For close to four billion years, the biological world has been honed by natural selection to handle, with great care, the available supply of energy and materials—to squeeze every drop of usable energy from every possible source and maximize the amount of life that can be supported by planet Earth.

Photosynthesis allows living organisms to harness sunlight, but life hasn't always used solar power. The biological cells that we are familiar with emerged early in the history of Earth from prebiotic systems—about which we know almost nothing, probably because their materials have long since been recycled! Since then, the process for managing energy within biological cells has been continuously evolving.

As organisms began to make use of sunlight, they were able to manage energy and materials more efficiently. The ecosystems they formed became more circular, with better exploitation of waste materials and re-use of pre-built molecules passing from organism to organism. To appreciate how energy and material management go hand in hand within circular systems, and how this contrasts with our own linear economic system, it helps to understand how the earliest life forms made a living.

Extracting energy from the environment

Right back at the beginning of Earth's biological history, single-celled organisms did not use photosynthesis. Instead they used a process called chemosynthesis, which basically translates as using batteries instead of solar panels. There was, and still is, a wealth of energy stored in the Earth's crust—energy bound within the molecules of the planet as it formed. That energy has been continuously resupplied by geological processes and nuclear fission, deep inside the Earth's core, ever since. The first living cells on Earth evolved to exploit geophysical energy by chemically reaching out into their environment and making electrical connections between inorganic materials like hydrogen sulfide and oxygen to create a current—a flow of electrons—between chemicals. Using the energy from these minerals, the bacteria powered their internal systems and performed functions like making sugar from carbon in their environment.

To perform this process, the first living organisms needed the ability to capture electrons in their surroundings using a chemical reaction called oxidation, and to do that they needed tools! The tools living organisms use for chemosynthesis are proteins—a versatile class of molecule that underpin almost every natural process, from cellular chemistry to signaling between cells. Proteins are polymers (long chains of repeating units, page 20), and they are constructed by stringing together molecules known as amino acids.

There is a multitude of ways to arrange an individual protein into a compact 3D structure, but only a single conformation where the chain is uncoiled. Therefore, thanks to Brownian motion, it's more likely that we'll find a protein scrunched up into a ball rather than stretched out like a necklace. The precise way in which a particular protein folds up is determined by the sequence of amino acids in the chain, which, in many cases, results in predictable three-dimensional protein structures for particular environmental conditions. This characteristic has made proteins one of the most powerful classes of molecules in the natural world and is the reason evolution has been able to shape their behavior so precisely.

All life forms deploy specialized proteins called enzymes to manipulate chemicals in the surrounding environment and extract the energy and raw materials they contain. An enzyme's 3D shape allows it to interact with the 3D structure of particular molecules, and in doing so, it can affect their conformation. We mentioned energy barriers in chapter 1, where we defined the energy barrier of a material as its ability to resist rearrangement. Enzymes are biological catalysts, meaning they lower the energy barrier for a particular chemical change; they make it easier to rearrange a molecule, thereby reducing the amount

A natural electronic circuit

Mitochondria

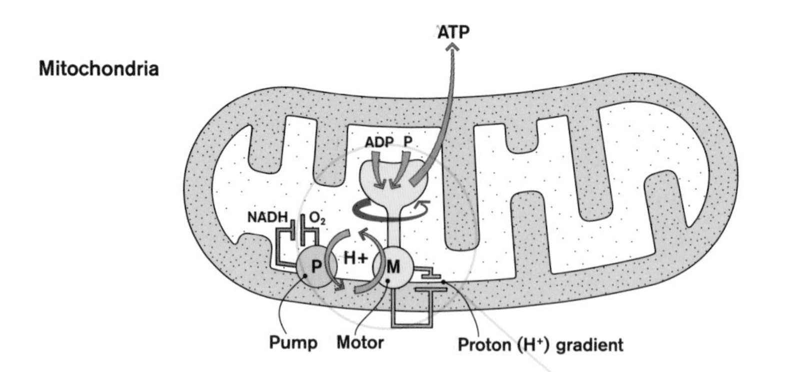

Electron transport chain

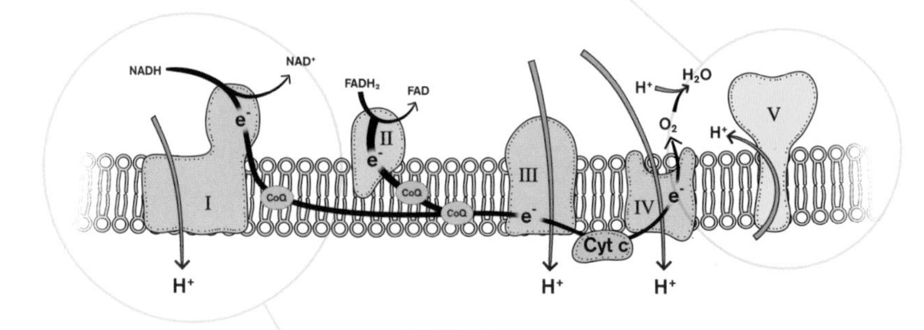

Respiratory complex

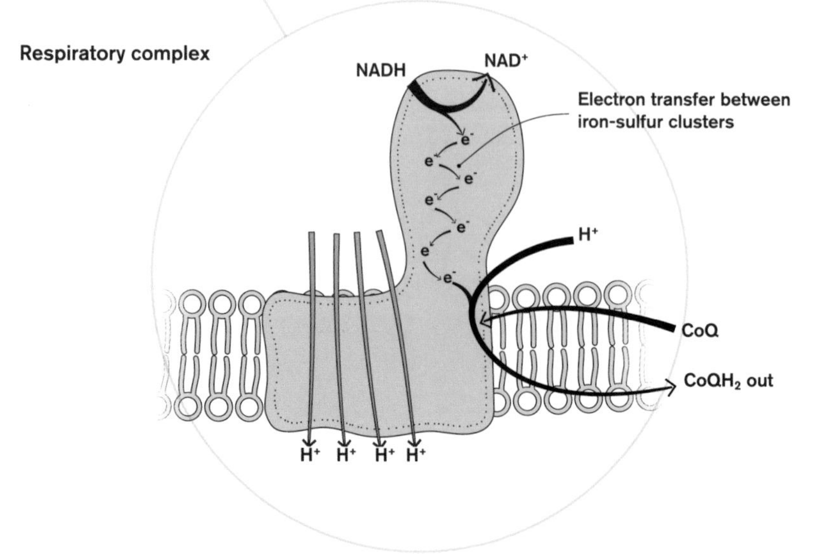

of energy needed to make a specific reaction occur. Incidentally, we also use enzymes, both industrially and in our own homes, to speed up chemical changes we want to perform. For example, if your washing detergent is "bio," it will contain enzymes like lipases and proteases to clean your clothes quickly, and rennet is the enzyme used by cheesemakers to convert milk protein into cheese.

A natural electronic circuit

During the first stage of chemosynthesis, biological organisms employ enzymes to oxidize chemical fuels in their environment, acquiring an energetic electron, and leaving behind spent fuel as waste. The energy of the electron is then transferred into the internal energy system of the cell where it is stored in a molecule called adenosine triphosphate (ATP). ATP is the universal currency of energy within cells, and it can be safely transported to every part of the cell by diffusion to power all kinds of activities. This centralized architecture means that processes within the cell can be powered regardless of the nature or quantity of fuel in the external environment. Every surviving biological system we know of has adapted to using ATP as a common energy carrier. For this reason, it is possible to transplant processes from one species into the cells of another and they will continue to function—a fact that genetic engineers, and biological organisms, routinely exploit.

With such a universal internal energy system in place, many single-celled organisms have evolved what might be considered the chemical equivalent of a Swiss Army Knife™—a multi-use toolkit of enzymes that allows them to extract electrons from a wide variety of chemical fuels in their environment. Suitably equipped, these organisms can adapt and extract energy in many different environments by activating the production of the correct enzyme—each time producing a specific waste product.

Opposite: Biological systems perform functions found in human electronics using molecular components. In mitochondria, molecular pumps use energy from chemical reactions (effectively a battery) to drive protons across a membrane that creates the driving force for a motor, which mechanically generates a molecule called ATP (top). Zooming in to the detail of the membrane, we see that an electron passes between different proteins while driving a proton pump (center). Electrons hop between molecules in proteins called cofactors, without losing energy (bottom).

The ability to oxidize a wide variety of fuels has enabled single-celled organisms to survive in some of the most extreme environments on Earth. Bacteria and archaea, both structurally similar single-celled organisms, are found in thermal springs, under glaciers and icecaps, buried deep underground, and in the vicinity of hydrothermal vents on the ocean floor. There are even reports of bacteria that have evolved to survive in a NASA clean room by consuming the organic compounds contained in the floor-cleaning products, even though those products were intended to kill them.

Similarly, in the two centuries since humans invented the first plastics, species of bacteria and fungi have adapted to consume them, although not at a rate that could make much of a dent in our current waste levels!

Compatible energy systems

The second stage of chemosynthesis, known as the Calvin cycle, uses energy from ATP to manufacture complex carbon compounds, like sugar, from CO_2 in the environment. Sugar can store energy for a long time, so that if fuel in the environment is diminished, the cell can call upon its reserves and use each sugar molecule to create up to 38 ATP molecules on demand.

Setting fire to sugar in oxygen releases the trapped energy very rapidly, but like the exploding stick of dynamite, this method channels the production of entropy very inefficiently. Instead, biological organisms have a remarkable mechanism—called respiration—for managing precisely the same chemical reaction, but without the entropy-wasting fire. Respiration occurs in a large cluster of proteins called the respiratory complex, which is an impressive example of biological electronics. The process of extracting the energy from stored sugar is very similar to the first stage of chemosynthesis; think of it as a roller coaster for electrons!

The electron from the fuel source is passed from protein to protein, like a subatomic baton in a biological relay race, down an electron transport chain, which is a series of molecules designed to move electrons around easily. At each step in the chain, energy from the electron is used to power a molecular pump that moves protons across a membrane. Protons are positively charged particles, so as they are pumped across the membrane, a charge difference builds up creating an electric field. This electric field temporarily stores alot of the energy from that original electron.

As the electron reaches the end of the chain it is handed to the terminal electron acceptor—a molecule such as oxygen, nitrate, or sulphate—to create another inorganic waste compound such as water, nitrogen, or hydrogen sulphide. This last step for the electron transfers a lot of energy to the growing proton gradient. Finally, the energy held in the proton gradient is used to drive a molecular motor—an enzyme called ATP synthase—whose rotation operates mechanical levers that force a phosphate molecule to bind with adenosine diphosphate (ADP) yielding ATP. This incredible process transfers energy from the original fuel source—whether that be sugar or a fuel from the environment—to be stored as ATP. The rest of the cell can use this energy easily—by splitting the single phosphate away again—which powers specific tasks and finally returns that stored energy to space as earthshine. Since the conversion between ATP and ADP is reversible, they can be recycled over and over again.

In short, biological systems use tiny amounts of energy one molecule at a time, selecting when and where earthshine is produced to power useful changes. This approach is in stark contrast to our own systems, which release large quantities of energy and create lots of entropy all at once, which may not make the best possible use of it.

Extracting energy from sunlight

During the first billion years of life on Earth, some bacteria evolved a new way to acquire energy: photosynthesis. Although there are many variations of this process, they all follow the same basic two-step approach as chemosynthesis. First, energy is captured from outside the cell—in this case from light—and temporarily stored in ATP. Then ATP is used to capture carbon from the atmosphere to build energy storage molecules.

In the first stage a photon hits a complex of proteins called photosytem II, which excites an electron, freeing it for use to generate ATP. In this reaction, a chlorophyll molecule has been oxidized, just like the inorganic fuels used in chemosynthesis. However, unlike chemosynthesis, the electron removed from chlorophyll is immediately replaced with a low-energy electron taken from a molecule of water, rapidly resetting the photosystem for the next photon.

Separating the electron source (water) from the energy source (light) gives photosynthesis a major advantage over chemosynthesis. The ubiquity of the input materials frees photosynthetic organisms from an eternal hunt for chemical fuels like hydrogen sulfide or ferrous ions, which could supply both the electron and the energy at

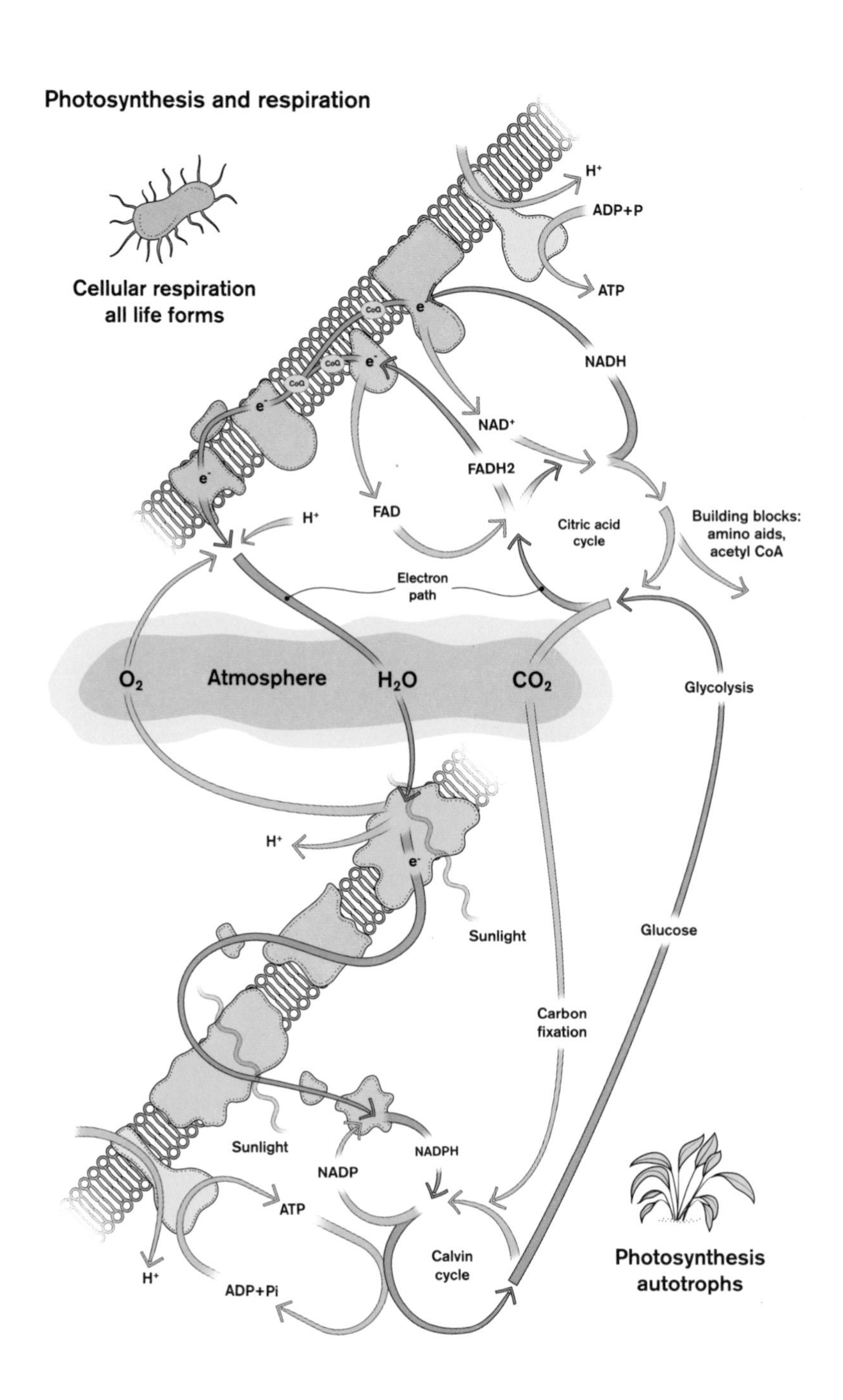

Photosynthesis and respiration
Cellular respiration
all life forms
H^+
ADP+P
ATP
CoQ
CoQ
CoQ
e
e^-
e^-
e^-
NADH
NAD^+
FADH2
FAD
H^+
Citric acid
cycle
Building blocks:
amino aids,
acetyl CoA
Electron
path
O_2
Atmosphere
H_2O
CO_2
Glycolysis
H^+
e^-
Sunlight
Glucose
Carbon
fixation
Sunlight
NADP
NADPH
ATP
H^+
ADP+Pi
Calvin
cycle
Photosynthesis
autotrophs

the same time, as required for chemosynthesis. Moreover, by not using a chemical fuel, there are no fuel remnants to become pollutants. The stroke of ingenuity was making the electron source for photosynthesis the same molecule as the waste product of respiration—water—*closing the material cycle*. This closed loop centered around water is at the heart of circularity in modern ecosystems.

After ATP is produced, the electron is passed to photosystem I, where it is excited by light a *second* time. Crucially, the electron itself keeps that energy and it becomes part of a molecule called NADP in a process called reduction—the opposite of oxidation.

The second stage of photosynthesis—the Calvin cycle—is extremely similar to the second stage of chemosynthesis. The energy from ATP and NADP is used to convert CO_2 into stable molecules—like fats and sugars—for long term energy storage. The electron—which started in a water molecule—is now stored in a sugar molecule, which is stable enough to be transported beyond the confines of a single cell, and can be stored for long periods of time as the complex branched sugar known as glycogen.

Biological electronics

Whether the original energy source is minerals in rock, sunlight, or the fluids emitted from a hydrothermal vent, living organisms have evolved to harness energy from the environment, and to do this they have had to develop biological electronic circuits.

Consequently, electronics were being deployed long before the ancestors of Faraday or Tesla walked on Earth, but the way biological organisms conduct electrons differs vastly from how humanmade electronics work. As we have seen, nature manipulates electricity at the single-molecule and single-electron scale, moving electrons from molecule to molecule in very carefully constructed biochemical chains.

The electron transport chains used in respiration and photosynthesis are molecular circuits; instead of wires, their electronics are constructed from chains of sophisticated electron-transport molecules called cofactors, which are encapsulated by certain

Opposite: Every organism in the ecosystem is part of an electronic circuit that is connected by the atmosphere. The two halves of the circuit are called photosynthesis and respiration. Electrons from water go all the way around the ecosystem taking the energy from the sun to make the building blocks of life, before returning to water to complete the circuit.

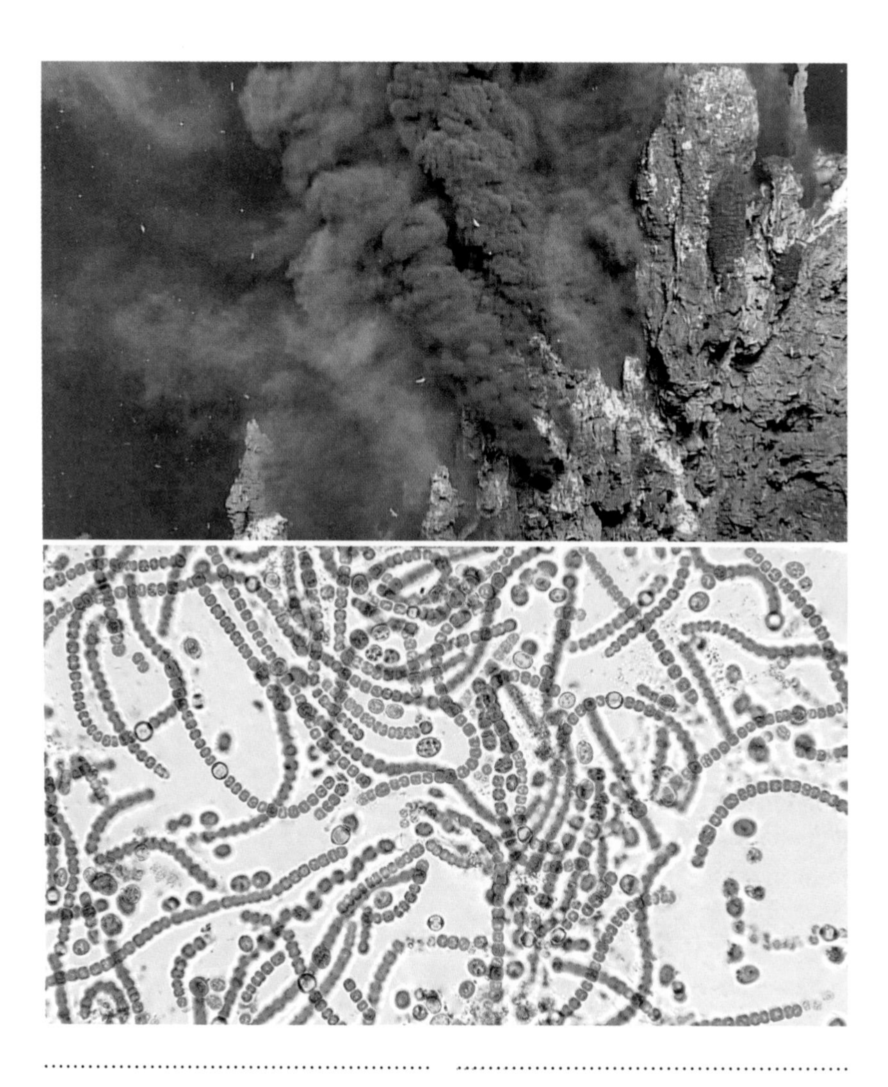

Above, top: Diverse communities driven by chemosynthesis are evident today. Deep-sea hydrothermal vents release chemical fuels from below the ocean floor and these chemicals can support entire ecosystems in the vicinity of the vent mouth. Courtesy of Schmidt Ocean Institute.

Above: Aquatic microbes such as these filamentous cyanobacteria were among the earliest photosynthetic organisms on Earth. They started harvesting solar energy around 3.5 billion years ago.

proteins and imbue the protein with additional abilities. The specificity of the 3D protein shapes controls which circuits connect to each other. Some cofactors contain a single metal atom in their center and are good at storing and releasing electrons. These cofactors are often brightly colored: heme, for example—the molecule that carries oxygen around your body and makes your blood red—changes color from dark to bright red when the iron cofactor at its center is oxidized. And it is chlorophyll—a bright green cofactor with a magnesium atom at its core, embedded in photosystems I and II—that intercepts red and blue sunlight on its way to space, making plants green!

What's more, the proteins containing cofactors can self-assemble into much larger circuits, giving organisms the ability to move electrons along chains of cofactors that link together the sites where carefully controlled chemical reactions take place. In this way, soft matter (page 20) can be used to construct conducting pathways via self-assembly, without having to generate the high-intensity energy necessary to manipulate metal wires into the right shape. That means that conducting pathways can be constructed using low-intensity energy sources. Photosynthesis is essentially a new circuit design that reuses some of the components involved in respiration. Once the ability to devise molecular electronics was mastered in the first billion years of life, it was simply a matter of rearranging the components differently that allowed cyanobacteria to start harvesting solar energy some 3.5 billion years ago. What circuits could we make in this very same way?

Mitochondria—a game changer for cellular electronics

In modern plant and animal cells, respiration occurs in specialized sub-units within each cell called mitochondria. Mitochondria are examples of cellular organelles—distinct modules within plant and animal cells that are surrounded by their own membrane. This internal membrane creates a barrier that controls the movement of molecules in and out of the gelatinous interior of the organelle. Indeed, the inner membrane in mitochondria houses the hydrogen pumps and molecular motors needed to create ATP, and it also represents an important stage in the evolution of life on Earth.

Evolutionary biologists believe that eukaryotes—all the plants, animals, and fungi we are familiar with, as well as a few single-celled species such as yeasts—evolved accidentally when one prokaryote failed to digest its lunch properly. The theory goes that one bacterium tried to eat another one, but instead it got "stuck." The smaller cell escaped its fate, became trapped, and evolved to live symbiotically inside.

Crucially, the first mitochondrion found a food source inside the larger cell and, because the mitochondrion was adapted to use oxygen, the host cell could now produce 38 ATPs per glucose molecule instead of 2 to 5. A good deal for both partners. Vitally, the trapped cell could reproduce alongside its new host and pass from generation to generation. Over the next half billion years, the two cells evolved symbiotically together to produce the mitochondria we know today. This symbiosis is how cells arose that possessed membranes inside membranes, and it opened the door to more efficient means of respiration that allowed eukaryotes to flourish and spread across the planet.

Chloroplasts are another type of membrane-bound organelle, which are crammed full of green chlorophyll, and are also thought to have arisen as a result of a meal gone wrong; an accidental symbiosis between a chemosynthetic bacterium and an early photosynthetic cyanobacterium that paved the way for algae and plants to evolve. The result is that we now have two symbiotic systems that are both operating in parallel to the mutual advantage of their host organisms. Between them mitochondria and choroplasts form a self-assembling energy-handling sub-system that can power entire ecosystems.

In the photosynthesis–respiration cycle, the electron starts off in water and ends at water. During the cycle, energy from the sun is transferred into sugar, where we can keep it stable for as long as we like, effectively stalling solar energy on its passage back into space. This observation is our first indication of how the principle of circularity can arise spontaneously from a set of common building blocks. The fuel and waste materials involved in one process can be the reciprocal waste and fuel for the second process: the output of one is the input of the other and vice versa. The whole dual cycle can be powered by sunlight and both halves of the process benefit from their continued cooperation. Furthermore, the combined effect of both halves of the cycle are responsible for the careful balance of gases—O_2, CO_2, and N_2—that make up our atmosphere, which is the common link between the two halves.

Evolved efficiency

Fundamental biochemical processes like photosynthesis have been around for billions of years, time enough for natural selection to identify the best, most efficient solutions for handling waste and processing fuel within the overall energy budget of the sun.

Renowned British naturalist Charles Darwin noticed that natural selection occurs spontaneously whenever four key conditions are met:

1. **Competition:** resource limits prevent every individual surviving to reproduce.
2. **Variation:** individuals within a population have different traits and characteristics.
3. **Heritability:** offspring inherit traits from their parents.
4. **Non-random survival**: heritable traits influence survival or reproduction.

Natural ecosystems fulfil all four criteria and as a result, traits that make individuals more likely to survive and have more offspring will tend to become more common in a population. Over time, these small changes in populations can add up to fundamentally change the characteristics of a species, or even create an entirely new one—this is evolution. We now know that offspring inherit traits from their parents in the form of genes—sequences of DNA that are turned into proteins in the cell and ultimately control the development, behavior, and appearance of the organism they belong to. As a result of natural selection, genes that make an organism well-adapted to its environment, increasing the number of offspring it produces—known as its genetic fitness—tend to become more common. Hence the term: "survival of the fittest."

No organism is an island

But no organism lives in isolation. Every individual is part of a network of organisms. Each environment is home to an ecological community, which, combined with the non-living components, such as water and rocks, forms an ecosystem. Within a given environment, any species whose food source runs out will ultimately become extinct, while new evolutionary variants that can exploit untapped energy sources—such as the waste of other organisms—enjoy a selective advantage. Over time, natural selection prunes the network so the species that survive are those able to replenish each other's food source. The result is an intertwined global system of recycling that operates efficiently within Earth's unbreakable energy boundaries.

A hectare of tropical rainforest may be home to thousands of miles of tree roots and branches and billions of leaves. In a single day, the forest's trees will pump hundreds of thousands of liters of water into the air and absorb tens of pounds of carbon dioxide from the atmosphere. In fact, the incredible matter-processing prowess of forests, is how the Amazon creates half its own rainfall and influences global weather patterns.

Dynamic ecosystems

Although no two ecosystems are alike—they vary as a result of climate, geology, altitude, and chance events—they are all governed by certain universal principles. The first principle is thermodynamics (page 18) whose laws state that matter and energy cannot be created or destroyed, and that entropy tends to increase. Ecosystems are also constrained by the law of stoichiometry, which states that chemical reactions must be balanced, and they are constrained by further principles of physics, such as the movement of gases and liquids, the mechanical properties of materials, aerodynamics, hydrodynamics, and others. The final principle is that the long-term fate of ecosystems is governed by unrelenting natural selection acting on the species within it.

Population geneticists often think of species in terms of gene pools, which for practical purposes we can think of as the catalog of all the gene variants being exchanged, over the generations, by a breeding population. Different variants of a gene are known as alleles and they compete, over evolutionary time, for superiority in the gene pool. But often, a population sustains multiple different variations of the same gene—no allele "wins" or "loses." This is why we have different hair or eye colors and wild animals of the same species often have different coats.

Constant variability within a species proves its worth when the environment changes, or when another species evolves or migrates into the environment, changing the dynamics of the ecosystem. Species with more variation in their population are more robust because it is more likely they will possess alleles that can cope with the altered conditions and offer a path for natural selection to follow. Crucially, success or failure of different genetic alleles depends heavily on other species within the ecosystem. For example, if a change in the environment causes numbers of a herbivore species to decline, their predators will be immediately affected, and so predators with alleles that enable them to prey on and digest a different food source may be favored by evolution.

Energy flows through ecosystems

The different functions performed by organisms within an ecosystem allow us to group them into layers known as trophic levels. Often these are visualized as a triangle or pyramid, which explains the feeding relationships that make up the food chain. The bottom, and by far the biggest layer of the pyramid represents the producers—organisms that can harness energy

Trophic levels

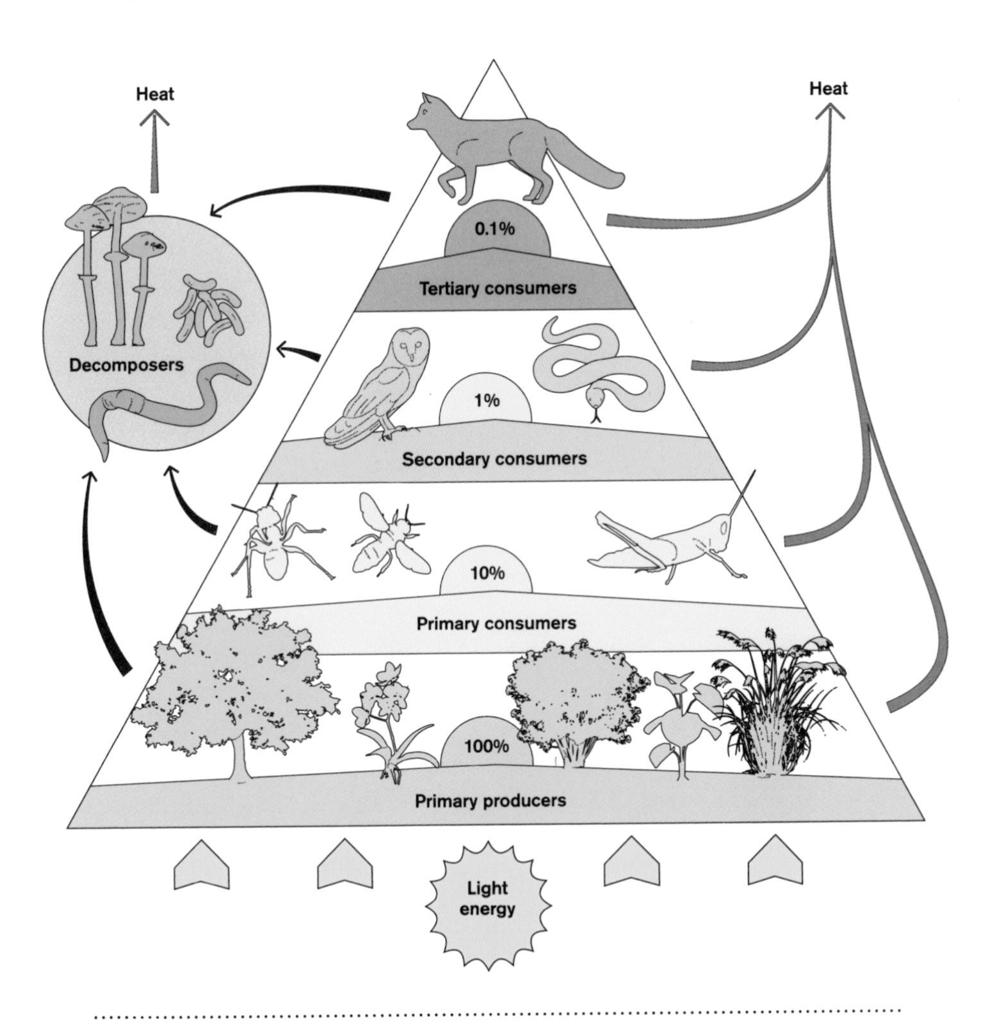

Above: Producers manufacture their own food from reservoirs of materials common to all organisms: soil, atmosphere, and bodies of water. Primary consumers are herbivores that feed on the vegetation, and are preyed on by carnivores called secondary consumers. Some ecosystems may support tertiary consumers, who feed on other carnivores, which are often the "top predators" of their environment. Finally, the material dance is completed by decomposers that recycle the remains of dead plants and animals back into natural reservoirs.

from the natural environment, most commonly the sun. They include plants, algae, phytoplankton, and other photosynthetic organisms, as well as chemosynthetic organisms, and the energy they produce supports the rest of the food chain. Apart from oxygen, water, carbon dioxide, and sunlight, producers like plants require nutrients to help them flourish. Nitrogen and phosphate are probably the two most important; both are used as artificial fertilizers for crops and garden plants. But wild plants must find their own sources of these key nutrients and have evolved different strategies.

Around 70 percent of land plants are thought to harbor soil fungi known as mycorrhizae in or on their roots. These symbiotic fungi extend a fine network of hyphae into the soil, which helps the plant to take up nutrients like nitrates. Some plants, such as the peas and beans, have evolved a symbiotic relationship with root-dwelling bacteria known as rhizobia, which are able to absorb nitrogen directly from the air.

The role of the fungal network in the life of plants is far more extensive than previously realized. The term "wood wide web" aptly refers to the incredibly complex network of fungi that extends beneath forests and woodlands allowing trees and plants to interact via chemical signaling. The trees provide nutrients for the fungi and in exchange the fungi communicate warning signals between trees and have even been shown to transmit immunity to disease around a forest.

The second trophic level identifies the primary consumers, the herbivores that feed on the producers. These organisms feed exclusively on vegetation, using the stored energy in the tissues of plants, algae, and other producers, to fuel their own metabolism, growth, and reproduction. Animal digestion breaks down food into water-soluble molecules that can be absorbed into the bloodstream and transported to where they are needed in the body. Carbohydrate and fat molecules can be used to produce ATP for immediate use, or stored for later, while protein molecules may be broken down into amino acids and re-assembled into new cellular proteins.

Herbivores are met with quite a challenge to extract the energy from the plant cells they eat. The thick outer wall of plant cells is made of cellulose, which most creatures are unable to break down. To access the juicy nutrients trapped inside the plant cells, many plant-eating creatures, including cattle, horses, goats, giraffes, deer, rabbits, and termites have evolved symbiotic relationships with communities of single-celled organisms—bacteria, protozoa, yeasts, and fungi—known as a microbiota, which break down cellulose for their host to digest.

The third trophic level is reserved for carnivores that feed on primary consumers. These are known as secondary consumers. At the fourth and, usually, final level are those organisms that feed on the secondary consumers. They are often referred to as top predators because it is rare for anything to prey on them. At each trophic level some energy is lost and by the top level there simply isn't enough energy left to support another layer of life. This is why the food chain is traditionally viewed as a pyramid—at each level less energy is available, therefore, in a balanced ecosystem, there are fewer herbivores than producers, fewer carnivores than herbivores, and top predators number fewest of all. This loss of energy constrains the shape and structure of ecosystems and determines a habitat's carrying capacity—the number of organisms that can be supported by the available resources. An ecosystem can only exceed its carrying capacity for a short period—unless resource-use efficiency changes to fit into the carrying

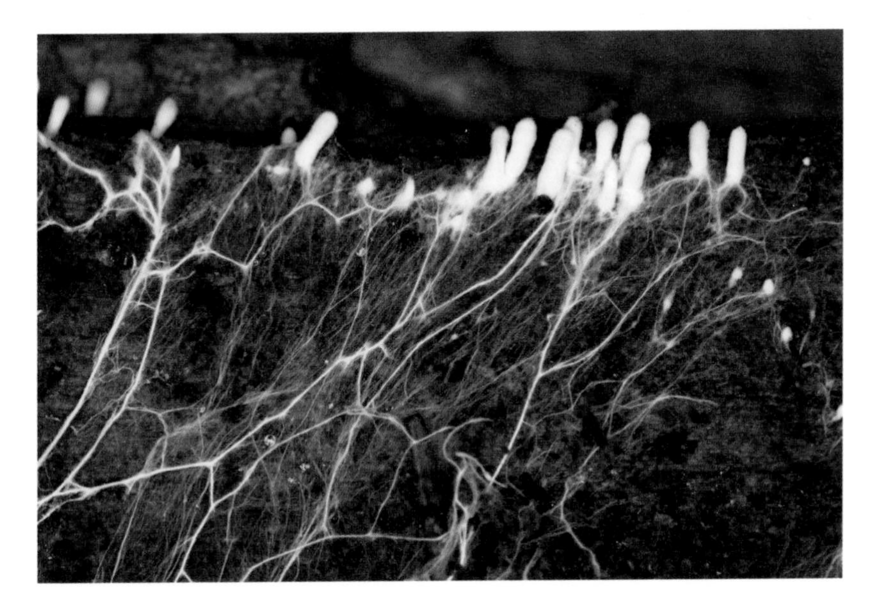

Above: Fungi are important decomposers in many ecosystems—their hair-thin networks of hyphae fill healthy soils and extract nutrients from decaying plant and animal matter, which is then returned to the food chain through symbiotic relationships with plants, or when fungal tissues are eaten by animals.

capacity, organisms will soon begin to die until the food chain falls back in line with the limitations of the environment. When organisms die, the energy and material stored in their tissues is recycled back into the ecosystem by decomposers, among which are bacteria, archaea, fungi, insects, and arachnids.

Although the trophic view of ecosystems has merits, ecologists have started using networks to describe relationships between species. Any ecosystem can be represented as a network of species, connected by ecological relationships: predators consume prey, decomposers recycle decaying matter, mutualists help their partners, and parasites feed on hosts. Within this network of organisms, it is clear that some species are better connected than others. Such central "nodes" of the network interact with many more species than those on the network's edges. These central species are known as keystone species because they are integral to the functioning of the ecosystem and any changes to their population will have a ripple effect throughout the network. Keystone species act as both a gauge and an emblem for the health of the entire ecosystem.

Loss of a keystone species has disastrous effects on ecosystems. Take the example of sea otters in the Pacific waters off the west coast of North America. Sea otters feed on sea urchins and in doing so keep their numbers in check. Sea urchins feed on giant kelp, itself a keystone species. Beginning around the turn of the 18th century, humans hunted sea otters for their fur, nearly eradicating the species by the early 1900s. As a result, sea urchin populations soared, and the great kelp forests began to disappear. Only through careful conservation starting in the late 20th century have sea otter populations started to rebound, bringing the sea urchins back in check and saving this unique ecosystem from what seemed to be an inevitable fate. Such is the power of a keystone species.

Ecosystems in balance

From the energy captured by the ecosystem's producers, a myriad of insects and arachnids, amphibians and reptiles, birds and mammals, are able to make a living. For instance, a hectare of tropical forest might contain a million leafcutter ants, harvesting two hundred pounds of leaf matter per year, making them the bane of farmers in the tropics, but providing essential material-reprocessing services for the ecosystem. Other invertebrates live in the soil and leaflitter, along with trillions of microbes, processing and recycling the ecosystem's waste matter. It is easy to marvel at the incredible diversity and

Above: A jararaca (*Bothrops jararaca*; carnivore), toco toucan (*Ramphastos toco*; herbivore), jabuticaba grape tree (*Plinia cauliflora*; producer), and an orange fungus (decomposer) represent four trophic levels in a tropical forest ecosystem in southern Brazil.

intricacy of a rainforest ecosystem, but the truth is that every ecosystem, even a scorched desert or frozen tundra, is a marvel of energy and material efficiency.

However, this kind of circular ecology isn't inevitable; species and whole ecosystems can (and do) drive themselves to extinction. Natural selection often favors selfish, uncooperative behavior, but under the right conditions, mutualisms can emerge and once established, they can be very stable. Similarly, collaboration among our own industries isn't inevitable, but it can emerge if we create the right conditions.

Circularity has emerged spontaneously in natural ecosystems because the fundamental biochemical processes that drive life—chemosynthesis, photosynthesis, and respiration—have evolved to make use of the same set of input and output molecules across species: carbon dioxide, water, oxygen, phosphates, and nitrates. It is no

Above: Kelp are keystone species that form great underwater forests all around the world, and are home to hundreds of species. The Marine Mammal Protection Act of 1972, followed by decades of careful conservation efforts—such as those led by Monterey Bay Aquarium, California, USA, whose Kelp forest exhibit is pictured—have helped restore this delicate ecosystem.

coincidence that these chemicals form a sub-set of the planetary boundary conditions (page 16). The compatibility of core molecules like protein, ATP, and glucose was born out of chemical necessity but has enabled ecosystems to flourish by forming complex, interconnected networks of energy and material transfer between systems.

Viewed from the perspective of physics, continuous reprocessing of nutrients in closed systems is only possible if the total entropy increases. Reorganizing molecules as they move from organism to organism, and between ecosystems, generates some material entropy but the majority is released as earthshine. We all constantly radiate earthshine into the environment and enzymes dictate where and when that occurs. The ability to use energy from sunlight to convert waste (reversibly) into fuel allows life to continue indefinitely on Earth.

That's not to say that every organism is perfectly adapted to its habitat, nor every ecosystem is in perfect balance. Natural selection is a continuous process and organisms are constantly evolving to be better adapted to their current environment, which is itself constantly changing—perhaps now more rapidly than ever. During times of rapid change, ecosystems scramble to find a new balance, and currently, ecosystems around the world are out of balance as they struggle to adjust to the demands made by a human population approaching 8 billion globally, and operating under a linear, rather than a circular economy. Rapid expansion of urban areas and changing land use by humans has left many creatures without a suitable habitat; intensive farming practices are stripping nutrients from the soil leaving it infertile and vulnerable to erosion; climate change is putting some species out of sync with the rest of their ecosystem.

As we will see in the next chapter, our linear take-make-waste manufacturing systems lack the sophisticated energy management of the natural systems we have discussed.

Power supply

We begin our biosmartphone thought experiment with one of the most crucial components—the battery.

If you have a smartphone, it probably uses a lithium ion battery. This type of battery stores energy by transferring ions from a positive electrode (anode) made of lithium-cobalt oxide, to a negative electrode (cathode), usually made of graphite. When switched on, the cathode slowly passes those ions back to the anode, generating an electrical charge. Among the most advanced lithium ion batteries currently available is the Tesla, Inc. Powerwall™, which is designed to store the charge generated by solar panels and can store up to 13.5 kilowatt-hours of energy—enough to power an entire home for nearly two days. However, a recent innovation from John Goodenough, inventor of the lithium ion battery, is the lithium glass battery, which promises three times the energy storage density.

Biofuel cells (BFCs), an alternative inspired by living organisms, are currently under development. Instead of storing energy by inserting ions into an electrode, BFCs store energy in the chemical bonds within molecules. They generate electricity by directly taking electrons from those bonds using enzymes. Such devices have ten times the energy density of a lithium ion battery.

BFCs have several problems though. For example, the enzymes that extract the electricity from the sugar don't have a long shelf-life, so the batteries require chemical refueling. To get around this problem, we could reprogram living cells to act as biobatteries and manufacture their own electricity-extracting enzymes on site. Such a concept is known as a microbial fuel cell (MFC). For example, researchers in Germany have engineered the well-known intestinal bacterium *Escherichia coli* (better known as *E. coli*) to convert the sugar glucose into electricity. Or perhaps our biosmartphone could make use of *Shewanella oneidensis*, a bacteria that can be fueled by many different materials and releases electricity as a by-product.

These bacteria and others such as *Geobacter sulferreducens*, can produce long, wire-like appendages, which scientists are emulating to develop biocircuits to carry electrical charge around, so perhaps our biosmartphone's circuitry will be made of protein, not metal wires. However, we don't necessarily need to create an electrical conduit to create a closed circuit. Our biosmartphone could use solar energy to continuously power an MFC to generate sugars from carbon dioxide in the air and an on-board supply of water, which would *diffuse* around the inner workings of our biosmartphone, giving up energy where it is needed and releasing water and carbon dioxide again.

As transistors get smaller, they approach the size of molecules. A steady stream of fuel molecules delivered by diffusion from a nearby MFC might be the ideal way of delivering energy to such a molecular transistor. Such self-assembling molecular electron transport chains could enable us to channel energy around our biosmartphone without a single metal wire, making the device much easier to recycle.

A biofuel cell

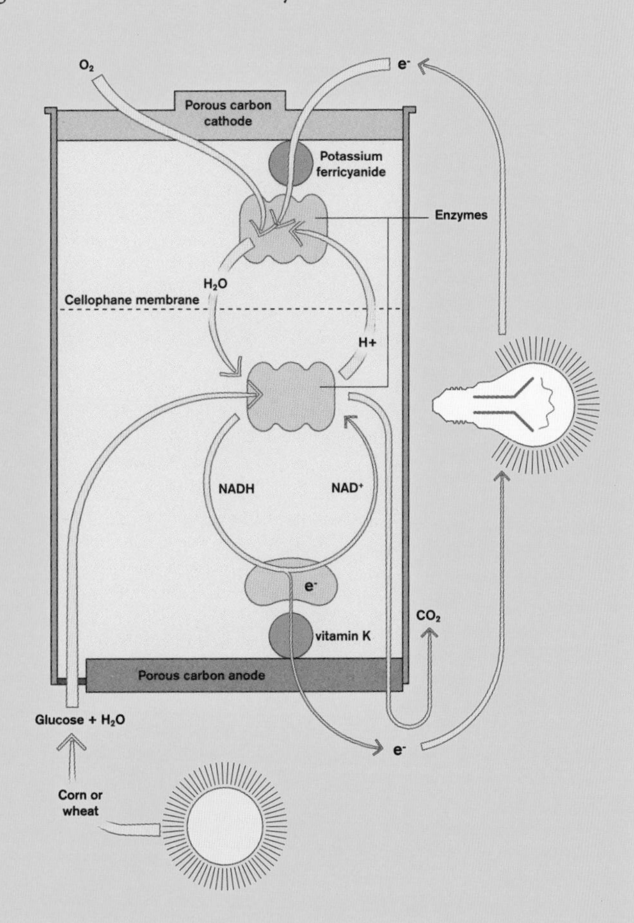

Right: A sugar-based biofuel cell uses enzymes to break down the sugar glucose into electrons, protons, and CO_2. The electrons are picked up by vitamin K and transferred to the anode, where they leave the battery as an electrical current. The protons combine with oxygen to form water and are released along with the CO_2—the only two waste products. The battery could potentially be "recharged" by adding more glucose, which can be sourced from crops such as corn and wheat.

3 Humanity's Linear Systems

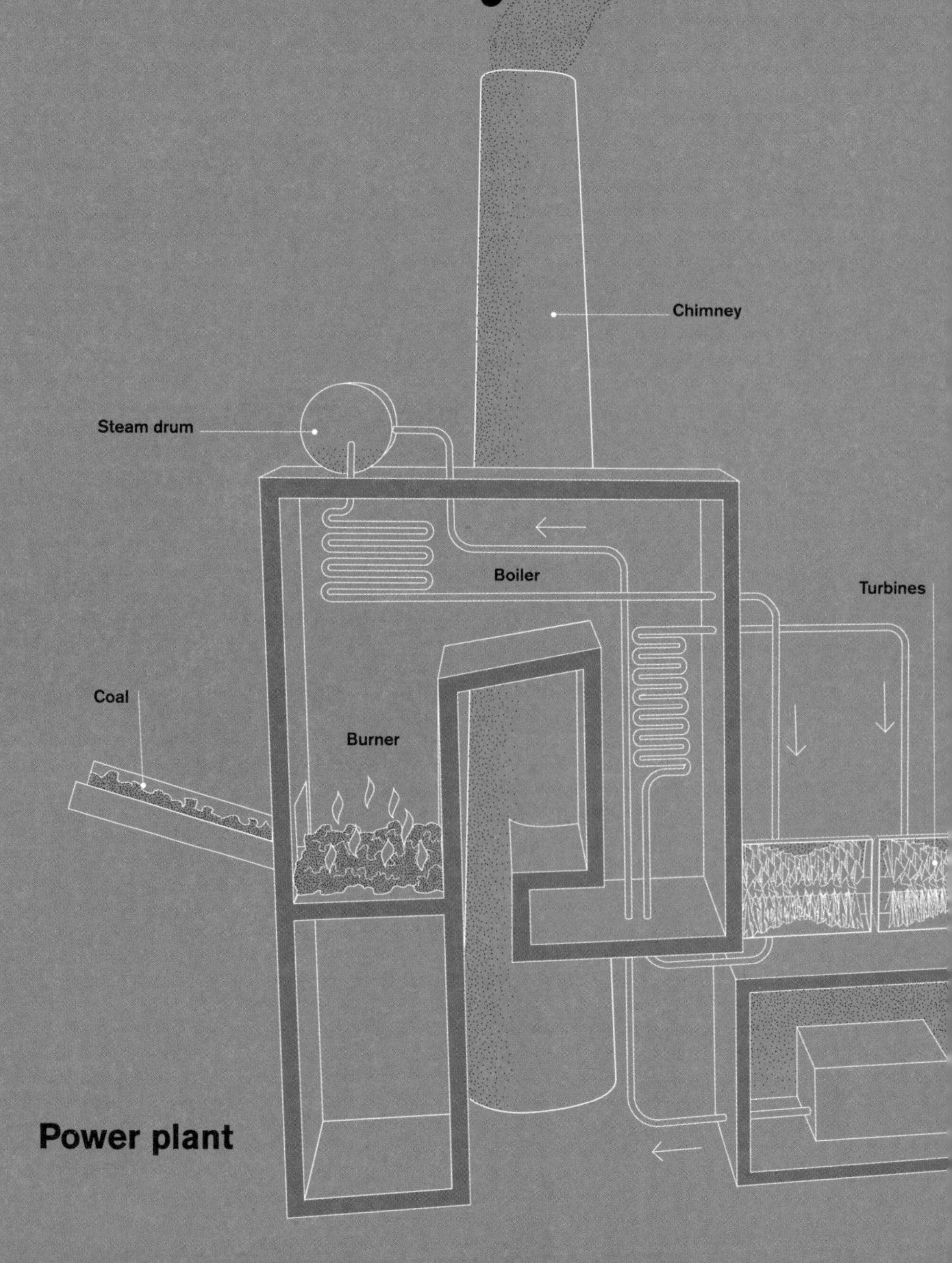

Power plant

“Life itself has existed for billions of years and has continually adapted to use materials effectively. It’s a complex system, but within it, there is no waste. Everything is metabolized. It’s not a linear economy at all, but circular. ...If we could build an economy that would use things rather than use them up, we could build a future that would really work in the long-term.”

Ellen MacArthur, March 2015. Founder of the Ellen MacArthur Foundation

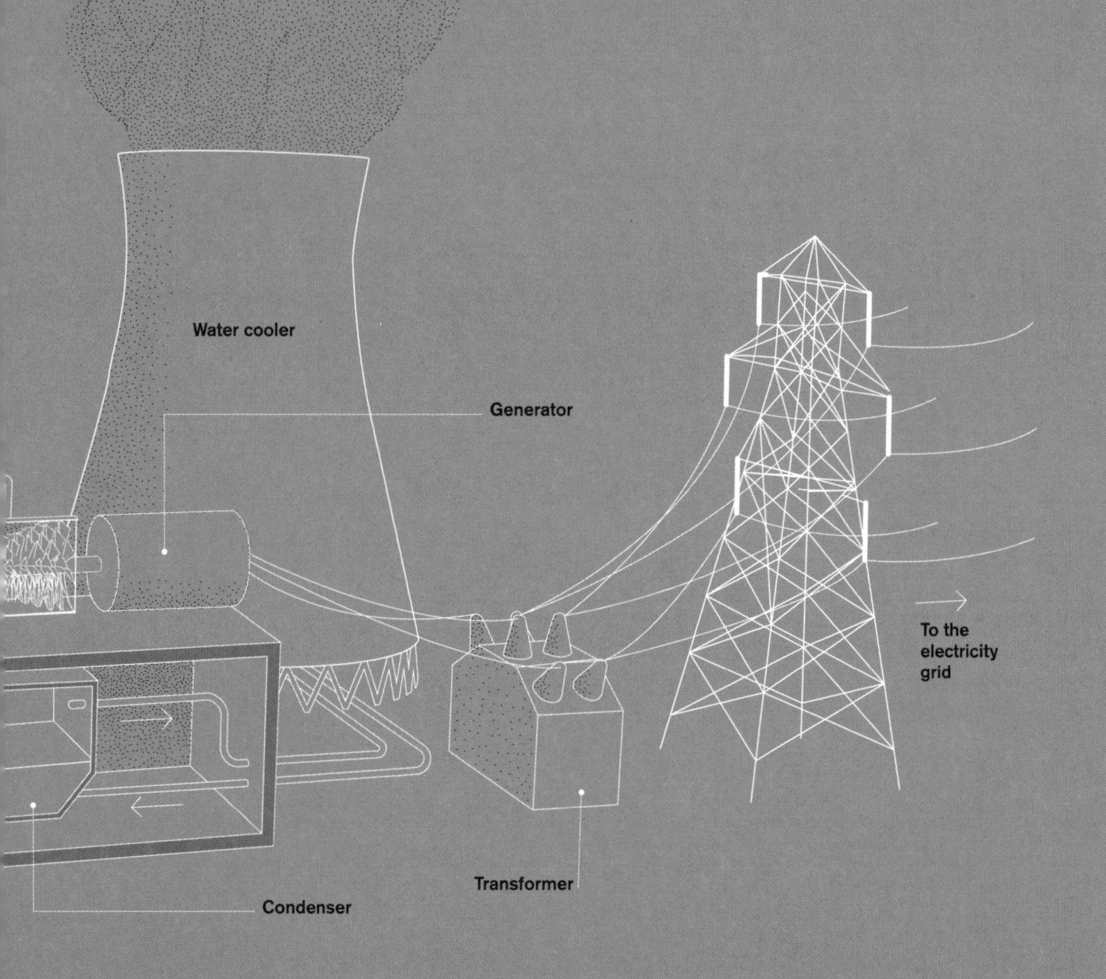

In contrast to the remarkable circularity of ecosystems, which arises spontaneously as a result of natural selection, our own systems for manufacture, distribution, and disposal are distinctly *linear* and rely on the assumption of infinite supplies of purified raw materials and energy, such as gold or gasoline.

From a day-to-day perspective, our systems appear to show remarkable innovation and efficiency. We can have almost anything we can imagine (and afford) delivered to our door in less than 24 hours. We can access any information we want at the drop of a hat from pretty much anywhere in the world. Ever-accelerating technological development blurs before our eyes but zoom out to the decadal timeframe, and what we actually have is an excellent system for filling garbage dumps.

The industrial revolution and 20th-century industries were powered almost entirely by fossil fuels and, just as the ecosystems that form around geothermal vents have adapted to their particular energy supply, so human industries and infrastructure adapted to the use of fossil fuels. Consequently, the transition to renewable energy sources, such as solar, wind, and hydropower, is neither swift nor simple.

Fueling the linear economy

Like the early lifeforms we discussed in the previous chapter, throughout human history we have opened up new opportunities by changing the fuel we use. When Stone Age humans discovered how to burn wood to create fire, this simple innovation led to a step change in our cultural evolution—keeping us warm, improving our diets, and powering humanity's activities for a million years. Similarly, the development of the first steam engines—which burned fossil fuels like coal to generate mechanical movement in a turbine—in the late 17th century was key to the industrial revolution.

Both wood and fossil fuels are a source of solar energy, trapped on its brief detour through Earth's atmosphere and stored in living cells. The difference is that while we can grow a tree and harvest the wood in a matter of decades, producing fossil fuels from dead plants and animals requires extreme temperatures and pressures, and a lot of time: about a million years, give or take. Slowly, under these extreme conditions, hydrocarbons—compounds made from hydrogen and carbon that once formed leaves and stems, muscles and skin—break apart to produce solid, liquid, or gas fuel. The final nature of the fuel product is determined by the ratio of carbon to hydrogen in the decomposing matter. Plants contain more carbon and tend to produce natural gases and coals, whereas animals, which contain more hydrogen, usually turn into oil.

These indirect sources of solar energy are high-intensity, meaning they release large, sudden bursts of energy when we burn them—more like the explosion of a stick of dynamite than a living creature metabolizing calories. These waves of energy are much larger than the Brownian waves that we discussed in chapter 1; they release vast amounts of energy (and entropy) in a very short space of time and can make big changes to the configuration of molecules as a result. Since we discovered these high-intensity fuels before we learned to harness solar power, our industries have adapted to make use of large bursts of energy to manipulate matter into useful forms.

Your wood-burning fire and an electrical power station both produce outputs that have the right chemistry to fit into natural global cycles of elements; in addition to electrical energy, power stations release mainly CO_2, water, and heat—the same outputs as cellular respiration. The problem is the ubiquity of these processes produces CO_2 more rapidly than the natural world can absorb through photosynthesis, exceeding the planetary boundaries. Why is the CO_2 output so high? Because our industries have standardized around materials with high energy barriers, such as steel, aluminum, copper, and glass, which must be softened by heating to make them malleable enough to work with. Furthermore, multiple waves of energy are applied to the same set of atoms as they move between different factories, first manipulating them into uniform blocks before reorganizing them again to produce useful products. Performing these high-energy processes in large, centralized power plants and factories allows us to exploit economies of scale, but it also results in lengthy and often complex transport networks for our energy and materials. These distributed supply chains are the cause of a great deal of waste.

For example, modern power stations use a steam-powered turbine to generate high-voltage electricity, which is then transported across a grid of electrical cables and passed through a series of transformers to reduce the voltage for industrial or domestic use. While such a method enables suppliers to meter and bill for electricity, during this journey up to 5 percent of the energy may be lost. Contrast that with a biological cell, where ATP can be delivered by diffusion without any loss of energy. This is because small Brownian fluctuations within the cell, which are caused by ambient heat from cellular metabolism, as well as the warmth of the sun, are enough to carry ATP wherever it is needed. Such Brownian energy transportation generally only takes place over the short distances within a cell, of course. By reducing the scale of our building blocks to the molecular, the size of the energy waves we would need would shrink to the point where sunlight could drive operations directly, eliminating the need to burn trapped solar energy in the form of wood or fossil fuels.

Operating over much shorter distances—biological cells range in size from about 1 to 100 micrometers (μm; millionths of a meter)—would create fresh scope for material efficiency through the integration of systems. Just as photosynthesis and respiration are implemented within individual plant cells as part of a much larger ecological pattern, matter could move between human energy and industrial processes at a molecular level, to create a global circular economy. And that focusses our attention from energy onto ways of organizing matter.

Subtractive and formative manufacturing techniques

While our skill may have increased, the basic principles of our two main manufacturing methods—known as subtractive and formative manufacturing—have remained unchanged since prehistoric times, and these processes require purified streams of raw materials.

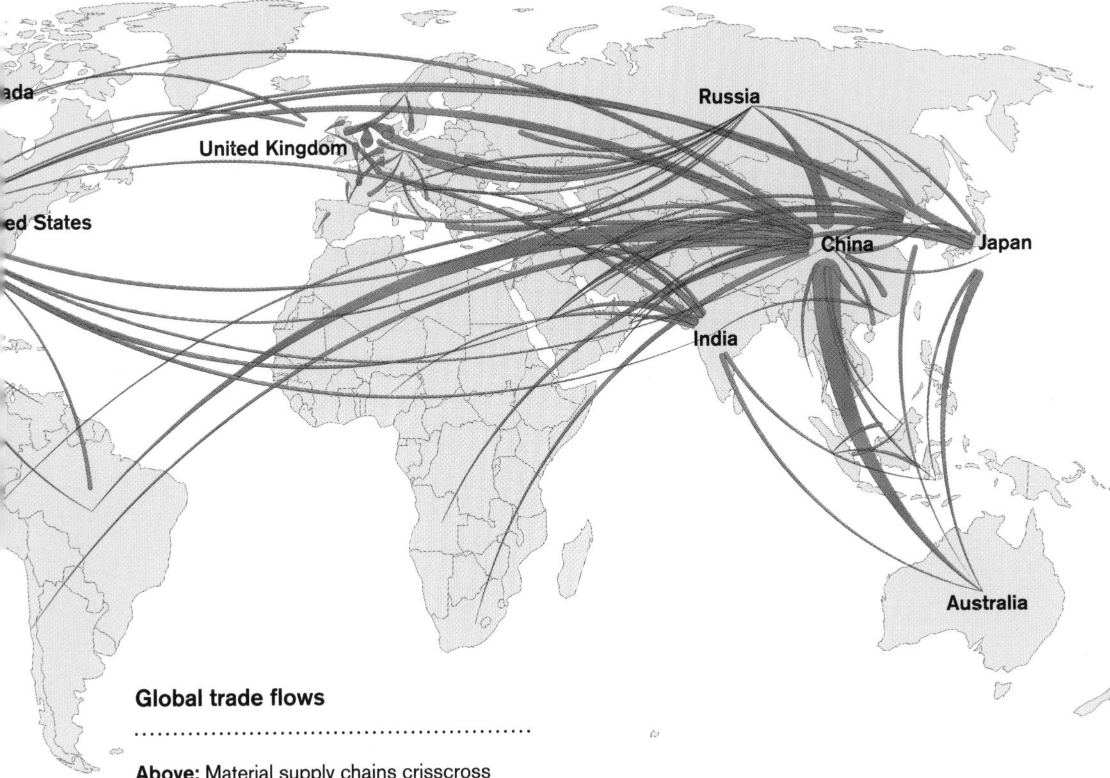

Global trade flows

Above: Material supply chains crisscross the globe, connecting centralized mining, refining, and manufacturing plants to customers. The flow of these materials around the planet is illustrated here with data from 2018, provided by Chatham House as part of the ResourceTrade.Earth project.

For more than two million years, early humans made functional tools and decorative objects by removing matter from a single, solid block of material. This is known as subtractive manufacturing: simple examples include knapping flint into an arrowhead or carving a block of wood into a bowl. Contemporary versions of these operations employ advanced computer-controlled milling systems or laser-cutting techniques that work on an extremely wide variety of materials, but the principle is the same.

The second main method is formative manufacturing. Here the mass of the work piece stays constant—nothing is removed or added—while it is re-formed into a given shape. The earliest known examples of firing clay to make pottery, which may have originated around 30,000 years ago, exploit a formative process, as do blacksmithing and glassblowing, and all three techniques are still in use today. Finally, subtractive and formative techniques can be used in combination, such as the lost wax process for making cast metal objects, which dates from 3700 BCE, and uses a clay sculpture as the negative for a mold to produce hollow metal objects.

Subtractive and formative manufacturing form the basis of the most advanced production processes. However, since the 1980s, an entirely new approach to fabrication has developed—known as additive manufacturing (AM)—of which the 3D printer is a great example. We explore the impact of AM on supply chains in the next chapter. But first we will consider two advanced techniques that can help to contextualize modern manufacturing.

Explosive forming

Certain highly specialized parts for very demanding applications, for example, rocket nozzles that must be very smooth, or spaceship bulkheads that cannot have weak points, require a method of construction called explosive forming. This is a brutal but ingenious technique designed to produce complex curved structures in a single work piece, avoiding weak points created by welds or changes in thickness or smoothness that result from other formative techniques, such as bending. Conventional formative techniques, such as pressing and rolling, can produce thin, flat sheets of metal with almost any level of tolerance for smoothness and thickness. Explosive forming takes such a flat sheet of metal and uses an explosive force to compress it against a die former—an object that has the right form, but created by conventional means, such as milling and polishing. The blast of explosives is effectively a perfectly symmetrical, spherical hammer, and the die is the anvil. As the form of the die is imprinted onto the metal object, the uniformity of the thickness and surface smoothness of the original sheet is well preserved on the formed piece.

Making a microchip

The violence of explosive forming contrasts with the delicate precision required for microchip fabrication. They are so ubiquitous we take them for granted, but the microchip is arguably the pinnacle of human manufacturing. The scale of their components—currently 5 nanometers (nm—or millionths of a millimeter)—and number (trillions) is rivaled only in biological systems. Their manufacture relies on a vast pyramid of human knowledge, from physics to chemistry, as well as engineering and commercial know-how.

Above: Subtractive techniques like knapping stone tools (top) and wood carving (center left), and formative techniques like pottery (center right) and glass blowing (bottom right) have been the backbone of human industry for many centuries.

Hundreds of chips are manufactured simultaneously across lattices of silicon crystals, known as wafers, which come in a range of sizes up to 12 inches (300 millimeters) wide. The process for creating such a gargantuan number of tiny features uniformly is nothing short of miraculous. The technique—known as photolithography—involves using patterned light to control where chemicals known as dopants, such as boron or phosphorus, are incorporated into the silicon lattice. Dopants are either electron-rich or electron-deficient, which affects the way electrons move through the lattice of silicon atoms and ultimately controls the functioning of the microchip.

The method exploits a phenomenon known as thin film interference—which is responsible for the rainbow of colors created when light reflects off a layer of oil on water. A beam of ultraviolet light (UV) is reflected from a stack of very thin, alternating layers of different materials and casts a complex pattern of bright and dark regions across a photosensitive coating on the silicon wafer. Information about the desired features of the final microchip are encoded in the UV reflection, causing chemical changes wherever UV light hits the photosensitive material. Dopants are then flooded across the entire wafer and penetrate the silicon only in regions that were exposed to UV. Repeating this process many times with different patterns and subtle variations means that a single 12-inch (300mm) wafer could, in principle, contain approximately 3,000 trillion distinctly doped regions—transistors—created in parallel in this incredibly efficient process.

Chip fabrication technology has been improving by a factor of two every 18 months for more than 40 years—a phenomenal rate of progress. Microchips with features just 2 nm in size are likely to be the final stage in the evolution of such a top-down manufacturing process, where the components are created all at once in situ from a single block of material (17,000 trillion transistors per wafer). To make electronics smaller, we could use molecular cofactors (electron transport molecules, see page 34) by making them in bulk and assembling them in bottom-up processes, mimicking biological systems. Scientists are already investigating such single-molecule approaches to electronics.

High-energy, high-waste industries

As we have seen, the majority of our subtractive and formative manufacturing processes require an input of *pure* raw materials. But when we use high-intensity energy to form simple blocks of pure material, we make poor use of our entropy budget.

Manufacturing a soft drink

Let's consider the chain of events that needs to happen for you to enjoy a soft drink. First, the drink itself needs to be manufactured—the raw ingredients are produced from the air, soil, and water by plants at various farms. The crops are then harvested and transported to factories where they can be refined into the processed ingredients that will go into your drink.

Sugarcane must be milled and boiled into cane sugar crystals, while flavorings like vanilla and cinnamon must be dried, before all the ingredients can be mixed together to form a syrup. If these ingredients are grown and manufactured abroad, then they will likely travel through a complex chain of factories, mills, bulking plants, and ports, before being shipped to their destination country and transported to the drink manufacturer for dilution.

Next, high-pressure carbon dioxide needs to be sourced and pumped in to add a bit of fizz. But before you add the fizz you'd better have cans ready to hold the drink, which now needs to be stored under pressure to prevent the trapped carbon dioxide escaping back into the air.

The cans and their ring-pulls will need to be manufactured at another factory, using aluminum ingots refined from the mineral ore bauxite, which of course first had to be mined from rocks. The ingots must be cast and then rolled into thin sheets (usually at another factory) before being molded into a can shape and transported to the drinks manufacturer to be filled and sealed.

Next, the filled cans must be coated with chemicals that contain marketing and sales information, such as barcodes and logos, then packed into cases, stacked into pallets, and loaded into trucks for transportation to the warehouse. From here, more trucks will carry the pallets to different shops, allowing you to walk in and buy a cool, fizzy drink on a hot afternoon.

And after you have enjoyed the beverage for maybe 30 minutes, and metabolized the energy it contains to power your body for about three hours, the discarded packaging has to go through an equally complex reverse process, involving yet more transportation between recycling or waste-processing centers, whereupon it ends up being recycled, or chucked into landfill.

The majority of our raw materials require mining or drilling, made possible by the bursts of high-intensity energy released when we burn fossil fuels. These processes produce ores—rocks that contain small flecks of useful materials such as gold, iron or aluminum—which must be refined to produce pure raw materials for manufacture. Similarly, materials harvested from crops often require extensive processing to remove unwanted parts of the plant—shells, husks, stalks, and so on—and produce finished textiles.

Refining ore can be a toxic endeavor requiring industrial-scale chemistry to separate useful materials. For example, large quantities of cyanide and sulfuric acid are used to dissolve gold and separate it from its mineral ore, and extracting rare-earth elements like cerium and neodymium—essential components for technology ranging from wind turbines to smartphones—uses strong acids and hazardous compounds such as ammonia and sulphates, and in the process can release radioactive elements, like thorium. Toxic compounds used during the separation of mixed-material ores are useful to generate pure inputs for manufacturing, but they become waste products at the end of the process.

New technologies are starting to emerge that can make these refining processes less energy-intensive and polluting. But these improvements—although welcome—still feed into a linear manufacturing system that takes pure raw materials and turns them into complex, mixed-material waste that ends up in landfill and the atmosphere.

Textiles are renewable in the short term, but they compete with our food crops for limited space on fertile land. Both activities have a significant impact on the nitrogen cycle—one of the nine indicators of planetary health. In fact, soil is another resource that we have been mining for decades, as we strip out nutrients like nitrogen and phosphorus faster than they can replenish themselves.

Our subtractive and formative manufacturing techniques need pure input materials, which results in dumping entropy into landfill as material waste *and* into the atmosphere via energy-related emissions, rather than emitting it to space as radiation. In the case of subtractive manufacturing, milling or cutting creates offcuts, sawdust, or metal filings. The energy used to make and transport those small flecks of material might end up being wasted if they cannot be recycled or reused in some way. In formative manufacturing, large amounts of energy are applied in short periods of time to soften material and overcome its internal energy barriers. Or if starting with soft materials like clay, subsequent steps such as firing are necessary to render the product useful.

You cannot build a computer without roads

To manufacture any product, whether that's a smartphone, a cotton t-shirt, or a rocket engine, the materials must be transported between every step of the manufacturing process. Almost all the products we buy and consume every day have been created through a centralized manufacturing process, where large refineries produce raw materials and factories mass produce the components that are then distributed to many different locations in turn, for assembly, retail, consumption, and waste processing.

To reap the benefits of mass production, we have become dependent upon centralized, linear processes, which inherently implies significant reliance on

Above: Metals like gold are often extracted using huge mines, such as the Fimiston Open Pit, known colloquially as the Super Pit, in Western Australia, pictured here, which extracts around 15 million metric tonnes of rock per year.

transportation—a weak link in the efficiency of our production systems. Between them, manufacturing and transport contribute around 35 percent of global greenhouse emissions.

Even a simple can of soft drink relies on an extensive manufacturing supply chain (see page 63). Usually, the more complex the product, the longer and more convoluted the process. As consumers, perhaps one of the scariest realizations is just how little we know about the detail of these supply chains. For example, agricultural products like soybeans, harvested at tens or even hundreds of different farms, are mixed together in processing plants before being transported to factories worldwide where they are transformed into commercial products such as soy sauce, edamame beans, tofu, and tempeh.

Poor documentation and lack of regulation in many regions and industries means nobody can tell you exactly where, or how the soybeans in your tofu, or the coffee beans in your latte, were produced. It's a similar story for countless other agricultural and mineral commodities from palm oil to gold. Very few industries can boast complete supply chain transparency from source to shop.

Such lack of detailed knowledge makes it impossible to monitor material cycles and their interaction with the natural cycles of elements. The opaque nature of modern supply chains is also a major barrier to consumers making informed choices about purchasing sustainable products. Transparent supply chains would be much easier to achieve if we moved away from a centralized approach to production, toward one where companies source materials locally. This is perhaps a more important consideration—from the perspective of sustainability—than whether a product is organic, free range, or genetically modified, for example. But without oversight and consumer pressure, it is extremely difficult to create the economic incentives for companies to invest in more localized systems of production.

Opposite: Waste electronic equipment (e-waste) is among the most difficult to recycle. The WEEE Man—a 23-feet (7-meter) statue constructed out of scrap electrical items, from washing machines to smartphones—was designed by Paul Bonomini and is displayed at the Eden Project in Cornwall, UK. The statue weighs 3.6 tons (3.3t), equivalent to the total amount of e-waste the average UK citizen throws away in a lifetime. The Eden Project, a popular visitors' attraction sited within a reclaimed china clay pit, is home to the world's largest indoor rainforest.

The supply chain of a 21st-century smartphone

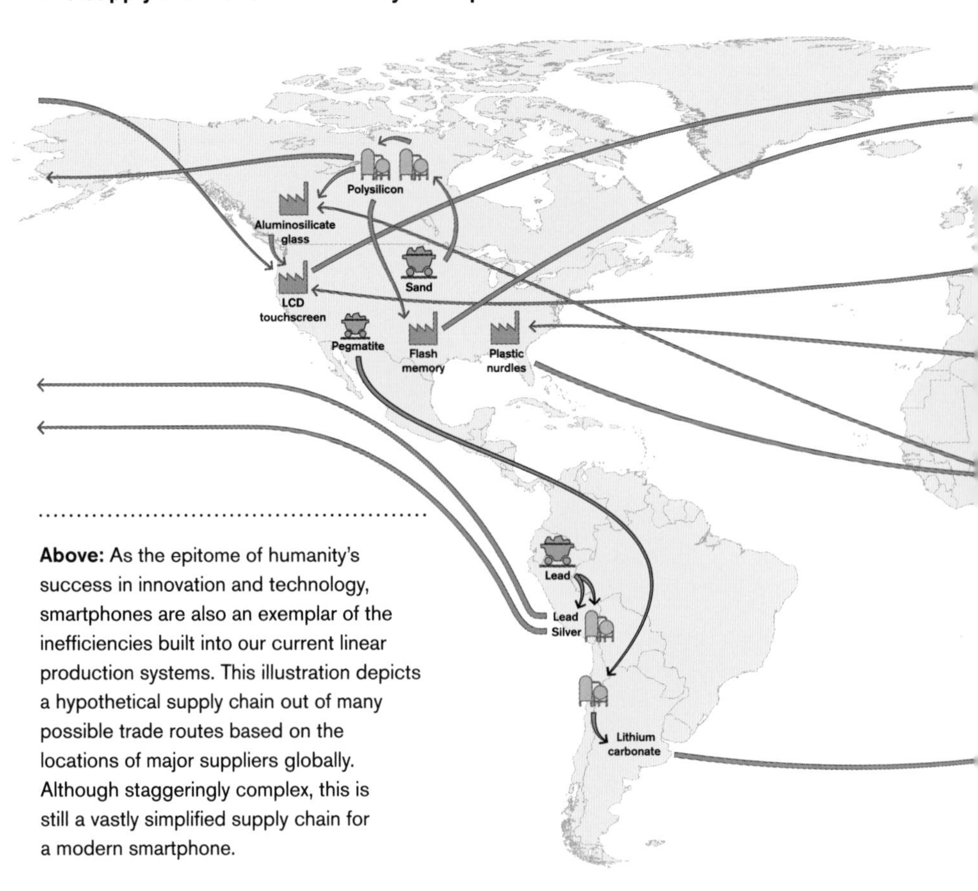

Above: As the epitome of humanity's success in innovation and technology, smartphones are also an exemplar of the inefficiencies built into our current linear production systems. This illustration depicts a hypothetical supply chain out of many possible trade routes based on the locations of major suppliers globally. Although staggeringly complex, this is still a vastly simplified supply chain for a modern smartphone.

Waste disposal

All too soon, products that require enormous amounts of energy and resources to manufacture are ready to be discarded, rendered obsolete by new technology. Although many of us try hard to recycle what we can, the odds are stacked against us, and the vast majority of everything humans make ends up either being buried in landfill sites or incinerated, both of which are major contributors of greenhouse gases.

Our electronic devices—everything from smartphones to domestic appliances—are a microcosm of this linear industrial strategy. Despite the enormous cost—both financial

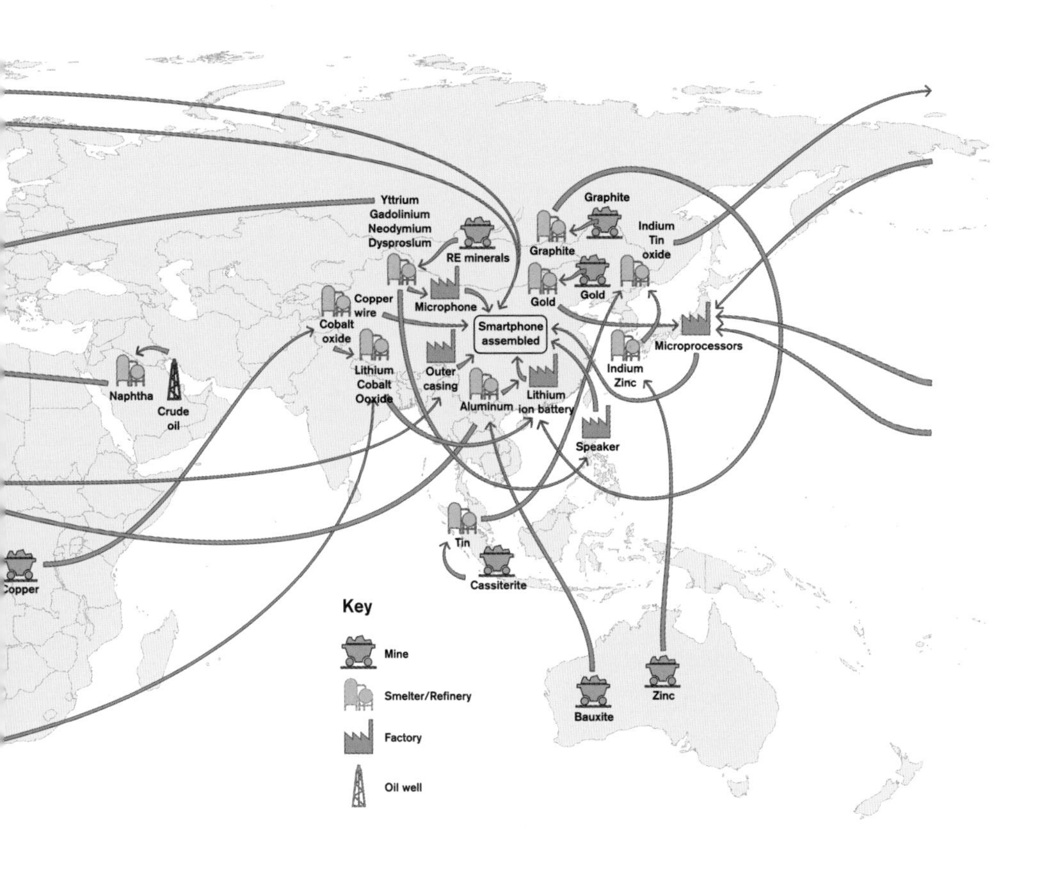

and environmental—of producing consumer electronics, just 17 percent of such "e-waste" is recycled. Part of the problem is that such complex pieces of technology require a complex configuration of materials, and the more complexity that goes into assembling the final product, the more difficult it tends to be to break those materials apart again. Although recycling and refurbishment schemes exist, we are producing an enormous amount of technological waste, whether it is left in a drawer or taken to a landfill site.

Sooner or later, the outputs of our linear manufacturing and waste systems all feed into the natural biogeochemical cycles of elements on Earth, such as carbon, hydrogen, and

Global carbon and nitrogen cycles

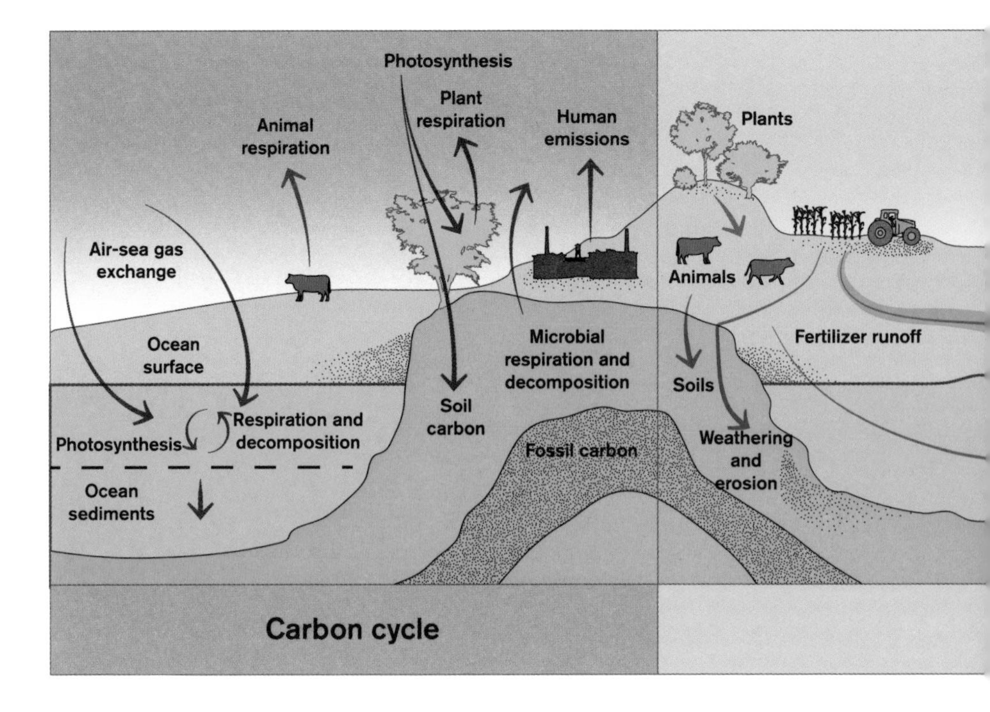

nitrogen. Every material on Earth is locked into an eternal cycle. Some elements, such as carbon and nitrogen, have extremely complex decades-long cycles; other cycles are simpler, their rates faster, for instance the water cycle. The waste elements we release back into the environment as we generate entropy in the pursuit of food, water, and energy, are added to the natural reservoirs of soil, water, and air upon which natural ecosystems depend.

From a single atom's perspective, our systems are nothing more than a detour on its eternal planetary circuit, but in manipulating matter on such an enormous global scale, we have fundamentally altered natural cycles. For example, harmful algal blooms can result when nitrogen and phosphate in fertilizers are washed into nearby rivers; water redirection to irrigate farms can cause water shortages elsewhere, and of course, atmospheric CO_2 levels have increased by 100 parts per million since 1958, linked to an average increase in global surface temperatures of nearly 1.8°F (1°C).

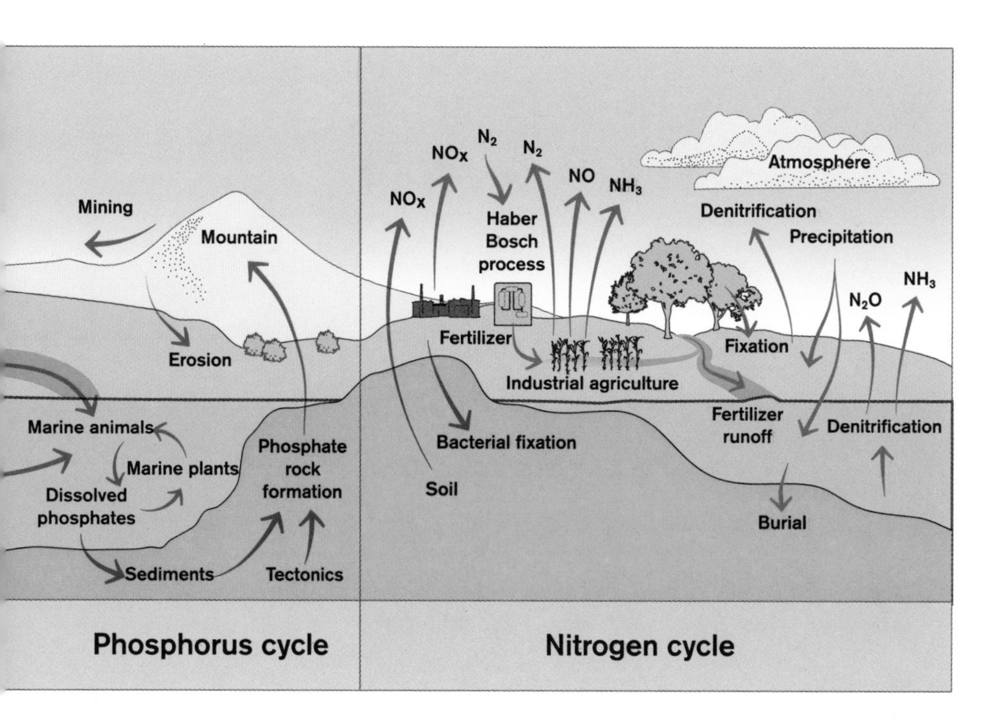

Above: Every element has its own cycle. Every year, 132 billion tons (120Gt) of CO_2 are taken up by photosynthesis and released via respiration. Water evaporates from rivers, lakes, and oceans, and falls again as over 120,000 cubic miles (500,000km^3) of rain and snow per year. Bacteria in the soil and inside plant roots can combine nitrogen with hydrogen to form ammonia, which is turned into nitrites and nitrates. Phosphate rocks naturally erode, and around 276 million tons (250Mt) are mined every year for use in fertilizers. Fertilizers are absorbed by plants or washed into rivers and streams. Phosphates from erosion, run-off, and decaying organisms are trapped in ocean sediments and eventually turn into rock.

The principles of a circular economy

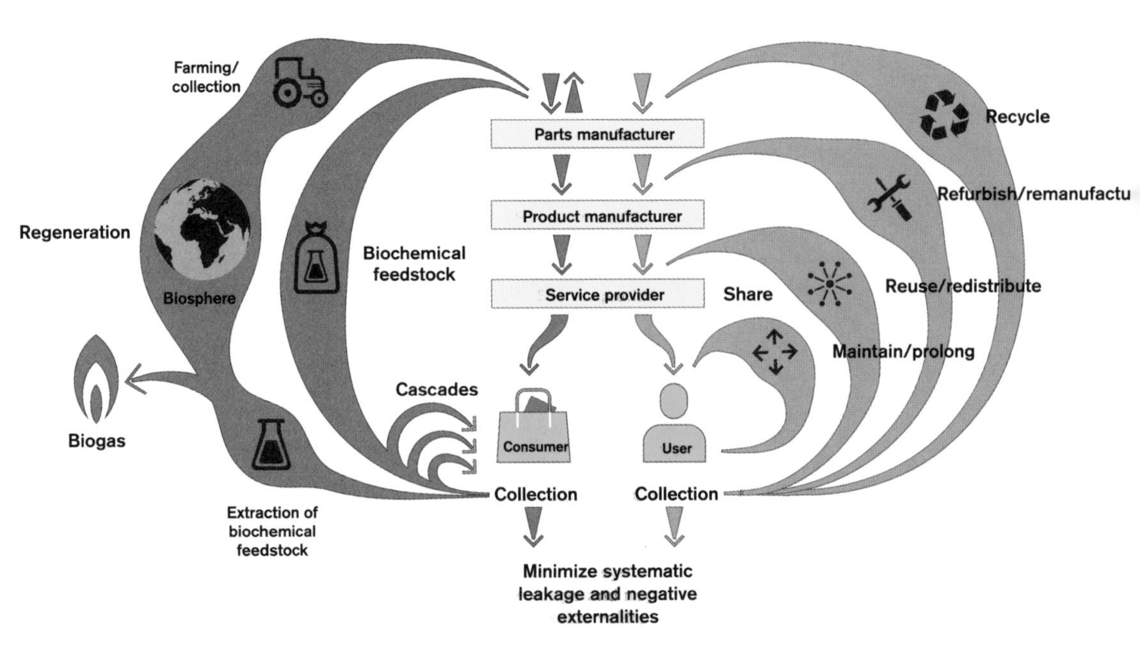

Above: A circular economy aims to rebuild and maintain five forms of capital—natural, human, manufactured, social and financial—through the continuous flow of materials by means of natural and artificial systems. In a circular economy, the objective is to create two flows of material, one technical (in blue) and one biological (in green)—which are never exhausted because they feed into each other in mutually supportive ways. With current technology, some technical materials cannot be recycled, so we aim to maximize their longevity and minimize their use. Maybe we can find technological replacements to design out unrecyclable materials.

Another way

Taking inspiration from the way natural ecosystems cycle resources and power their activities from low-intensity renewable energy, we can start to imagine an alternative to our destructive linear systems. As a result of natural selection, the organisms that successfully populate the world tend to be those that can efficiently replicate themselves using ubiquitously available sources of material, such as air, water or soil, and then at the end of their life, can be broken down to regenerate the reservoirs from which they came. This concept is also the foundation of a circular economy.

Consider the exquisite balance of materials found in soil, water, and air. These gargantuan reservoirs—the remains of countless generations and species—are not homogeneous, but they provide an auto-regenerative bank of useful raw materials that we can exploit. Most importantly, these reservoirs are within molecular reach of all living organisms, thus eliminating the need for elaborate transportation systems.

In a similar vein, the contents of landfill sites are not homogeneous; they represent a supply of the chemicals and materials that people need. Just as plants are able to extract nutrients from the mixed media of air, soil, and water, perhaps we could extract the materials we require from landfill. This prospect remains out of reach because our current manufacturing processes are not set up to use landfill as a resource, but if we could instead generate products that, at the end of their useful life, regenerated a reservoir of reusable materials like soil, the concept of "waste" would become obsolete.

Instead, chemical elements could be separated into one of two loops: a biological loop for biomaterials and a technical loop for technical materials. Wherever possible these loops are kept separate. They each contain a finite quantity of material, but through a process of re-use and recycling the materials circulate indefinitely to satisfy our never-ending, yet ever-changing, demands. At present, not all technical materials can be recycled.

Although such a circular economy contrasts dramatically with our current linear systems, we could successfully transition once the necessary changes to our manufacturing and energy systems are made. We need to carefully select a standardized set of material components that can be passed from one industry to another without generating waste products. Some modern industries are already putting this approach into practice, for example, S.Café® polyester is a fast-drying clothing material manufactured from recycled coffee grounds, and Parblex™ plastics make use of waste potato peelings to make durable alternatives to plastic packaging. Our goal must be to end built-in obsolescence and design products destined for a long life, that can be easily broken into their component materials. Finally, we need to localize our industries so that the waste they produce can be more easily integrated into an industrial symbiosis.

Although technological changes are needed for such a vision to become reality, the fundamental shift required is one of mindset—to view everything we produce as a useful material for another application, not as waste or a byproduct. Perhaps the biggest problem with the way that we make things now is that we have this concept of "the end object." Such thinking inevitably results in linear processes to create this desired object and what happens after its life has ended is often relegated to an after-thought. We will have to be much smarter than that with our materials if we are to transition to a circular economy.

Circularity by design

The Wissington sugar factory in Norfolk in the UK shows what can happen when a circular economy mindset is applied to a conventional manufacturing process. The three inputs into this remarkable factory are: sugar beets—grown on farms within 28 miles (45km) of the refinery—fossil fuel, and limestone. Using a clever, highly integrated design, the factory outputs a wide, perhaps surprising, range of products including refined sugar, electricity, lime, animal feed, biobutanol, medicinal cannabis, topsoil, gravel, and liquid carbon dioxide (CO_2). Despite burning fossil fuel, it also has a negative carbon footprint—its net atmospheric carbon emissions are negative, because CO_2 emissions from the power plant are piped through a greenhouse of cannabis, where the plants absorb much of the CO_2 to fuel photosynthesis, and use that energy to produce, among other things, the medical compound cannabidiol (CBD). Of the 3.9 million tons (3.5Mt) of raw material entering the refinery per year, fewer than 110 tons (100t) goes to landfill, and most of that is from the canteen! The factory is clearly profitable and demonstrates the great commercial opportunity that exists for companies with the imagination and capacity to adopt a circular mindset.

Mutualistic industries

The brutal, competitive side of nature is often portrayed in natural history documentaries and films. We are used to witnessing predation, parasitism, competition, and deception. A closer look reveals the extraordinary level of cooperation too. When organisms work together harmoniously, their relationships are described as mutualistic: think of bees and the plants they pollinate, or the microbes that populate the digestive systems of many herbivores. There are many and varied forms of mutualism in nature: ants tend large herds of aphids, protecting them from predators in exchange for a chance to feast on their sugary excrement; cleaner fish consume dead skin and parasites from larger sea creatures including sharks—these are examples of organisms making use of each other's waste products and putting them to good use.

If we similarly think of industrial and manufacturing processes as different "species" in a larger ecosystem, we can imagine many mutualistic relationships that could arise. Perhaps, in the future, every one of our industrial "species" will be able to live in mutualistic symbiosis with one another, passing waste materials from one company to another in a perpetual cycle that eliminates the need to mine scarce minerals and avoids polluting the environment without compromising financial growth.

The world's first such industrial ecology, Kalundborg Symbiosis in Denmark, offers a glimpse of what the future might be. This industrial park near Copenhagen is currently a partnership between 11 companies. First established in 1972, the collective operates on the principle that any waste or residue produced by one company must become a resource for another, ultimately benefiting both the economy and the environment. In total, the symbiosis involves more than 30 different material exchanges.

At its center is Asnæs Power Station, a coal-fired power station producing 1500 megawatts (MW) of electricity and passing on enough surplus heat energy to warm 3,500 local homes and a fish farm. The power station also produces excess steam, which they trade with an oil refinery in exchange for waste gas, which can be used to create more steam and more electricity (reducing their use of coal). Steam is also sold on for use by a pharmaceutical company, Novo Nordisk, which produces around half of the world's supply of synthetic insulin, used to treat diabetes. Other waste products from the power plant go into industrial enzyme and acid manufacture, cement production, and road building. The industrial park produces very little pollution because everything is cycled back into another industrial process before it has a chance to enter the wider environment. Kalundborg also uses very little water, because much of the wastewater from one industry is passed to another for use again.

Like biological symbiosis, this partnership was not directed from above—there was no top-down government directive or local incentive scheme—it arose spontaneously as a result of collaboration between companies that understood the value in a circular approach.

Industrial symbiosis

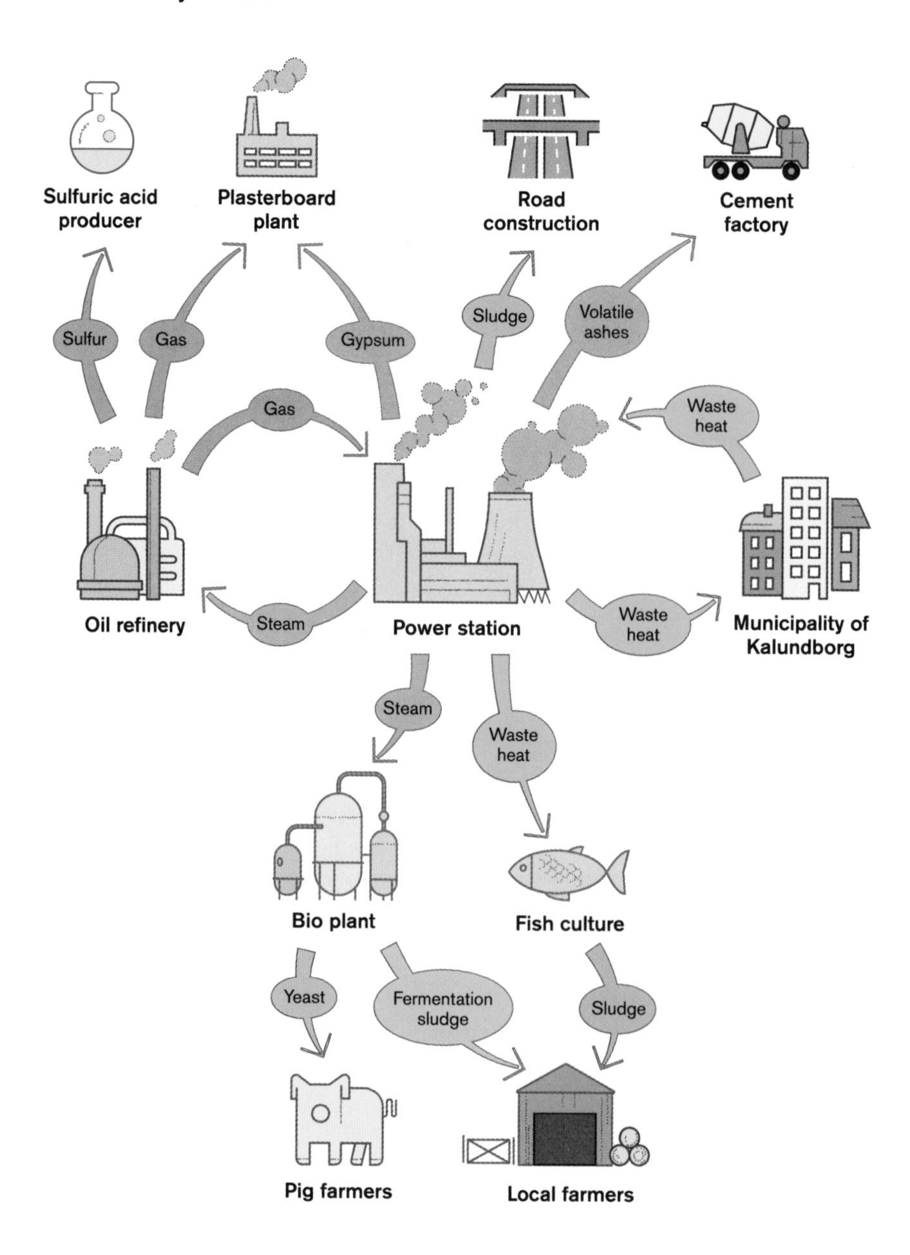

Giving circularity a helping hand

Although Kalundborg offers a tantalizing glimpse of a future circular economy, not every modern industrial process can fit so neatly into an industrial symbiosis, and the cost of transformation may impede some industries. Perhaps we require some truly transformative technological tools to help companies find ways to circularize?

Indeed, two such powerful technologies have been steadily emerging over the last three or four decades: additive manufacturing and synthetic biology. AM was already mentioned and is an exciting and active area of research that we will explore in the next chapter. Synthetic biology makes it possible to reprogram the DNA involved in cellular machinery to give us complete control over the chemical tasks that biological cells can perform on our behalf—from synthesizing drugs to controlling industrial reactions—with almost the same ease as programming a computer.

Both technologies lend themselves to creating new material building blocks, which like natural materials, can be fabricated and manipulated using low-intensity energy sources, with knock-on benefits for manufacturing, transport, and waste transfer between companies. Together these technologies could give us complete computational command over the entire manufacturing process from the molecular to the macro scale. Such a change would greatly simplify the creation of industrial ecosystems. With additive manufacturing and synthetic biology in our toolkit, the economic mindsets embodied by Wissington and Kalundborg may become the norm, rather than the exception.

Opposite: Kalundborg eco-industrial park operates with very little waste. Industrial partners trade waste materials to produce useful commodities, so virtually all inputs become revenue-generating outputs. The power plant produces excess steam that can be used by the oil refinery, which produces waste gases that can be used in the production of plasterboard; the fish farm uses waste heat from the power plant and produces pond sludge that can be used by local farmers for fertilizer, and so on. This innovative network won the park the Gothenburg Sustainability award in 2018.

Sensory capabilities

To build a biodegradable smartphone we need to find soft, or easily recycled materials that can perform the functions our phone needs. We must be able to send and receive electromagnetic, mechanical, and chemical signals and convert them to and from a standard internal signal. Fortunately, the natural world offers a rich catalog of sensors made from soft materials that we could tap into for manufacturing inspiration in a circular economy.

Light-sensitive molecules—such as those found in the eyes of creatures as diverse as humans and spiders—convert visual information into chemical signals and could form the basis of a biosmartphone camera. Conversely, light-emitting molecules responsible for the bioluminescence of glow worms or jellyfish could output visual information in response to directions from the user, such as opening an app or clicking on a photo. There are also many strategies for manipulating light, such as the structural color systems found in butterflies, birds, chameleons, octopuses, and squid. Such structures can control reflections in the ultraviolet, visible, and near infrared color ranges, and enable complex behaviors such as camouflage or mating displays.

Touch sensitivity—an important component of a smartphone screen—is a complex chemo-mechanical response. Skin combines pressure-, vibration-, itch-, temperature-, and pain-receptors just below the surface to translate physical information into an electrochemical impulse.

Our microphone might use tiny hairs—like the cilia in human ears or the trigger hairs of Venus fly traps—to detect mechanical vibrations. Sounds can be produced by actively vibrating membranes—as with the deafening call of cicadas—or by forcing air over resonant structures such as our larynx or the bony syrinx found in birds. The remarkable vocal dexterity of species like the superb lyrebird (*Menura novaehollandiae*)—rumoured to be able to mimic any sound—teaches us that biological systems can replicate diverse and complex sounds, and hint at the sound quality our biosmartphone might be able to produce.

Perhaps most importantly, our biosmartphone would need to generate and detect radio waves. While certain sharks can detect electric fields and some birds and mammals can sense magnetic fields, no living organism has been discovered that exploits modulated radio waves as we do, and whether biomolecules can interact via radio frequencies is a subject of active research. Perhaps biological radio communications never evolved because simpler, energetically cheaper, effective communication systems exist; whales, lions, and elephants can communicate over large distances using low-frequency infrasound. But that doesn't mean biological radio frequency modulation couldn't be possible, with a bit of clever bioengineering.

Our phone would also need to have a sense of where it was in the world, detecting GPS signals and using motion and gravity to determine its orientation. Plant roots sense gravity using dense, starchy structures called statocytes, which settle to the bottom of cells and co-ordinate the movement of growth hormones, cell by cell, toward the bottom of the plant. Similarly, the human vestibular system determines our motion and orientation from the movement of calcium carbonate molecules in the fluid-filled canals of the inner ear.

Chemical detection—smell and taste—could add extra functionality to our biosmartphone; imagine a phone that can warn you about a gas leak or check your breath for markers of disease. Smell and taste rely on chemical receptors on the surface of cells that bind physically to one or several target molecules and trigger nervous impulses.

Left: This electron microscope image of a fruit fly's (*Drosophila melanogaster*) tarsus (foot) shows it to be a marvel of sensor integration. The tarsus supports a myriad of chemical and mechanical sensors all integrated into a region smaller than the tip of a human hair.

4 Additive Manufacturing

"The understanding of complexity and the use of the creativity of nature, the continuation of the work of nature are the grand challenges for the scientists of the 21st century."

Ilya Prigogine, *Is Future Given?*, 2003

3D printer

The previous chapters have explored the foundational relationships between four cornerstones of the human world. The roles played by energy, materials, manufacturing, and transport were examined in both human and biological contexts, and at planetary and molecular scales. Our whirlwind discussion offered the view that a circular economy is greatly preferential for humanity compared to the current linear system, and these arguments were couched mostly in terms of planetary physics and biological precedent, as well as the obvious economic unsustainability of linear flows.

As we saw in the last chapter, in a circular economy the principal objective is to generate two flows of material—one technical and one biological—which are finite in capacity but never exhausted. Such a scenario is only possible by closing the loops of material flow—in other words—recycling *everything*. This regenerative ability is the very property of our natural systems that the linear economy undermines.

Once such circular flows of material are fully stocked with resources mined from the ground, there is no additional value to be gained by increasing the quantity of material in circulation. Instead, value arises by embedding new information into existing materials through their constant upgrade, repair, or remanufacture.

And so, a fifth foundational component comes into the discussion: information. In this chapter, we explore the role that information plays in biological and manufacturing contexts, enabling a different kind of manufacturing, one that underpins biological ecosystems and could be the foundation of future circular technologies.

What is information?

Information is a word we use all the time, but is surprisingly slippery to define generally. The definition of information has occupied philosophers and mathematicians for centuries. The race to understand, define, and control information has ended wars and birthed new technology like communications, computing, and artificial intelligence.

If entropy (page 18) is a measure of the number of ways we can organize a system then information is the particular configuration the system adopts, and changing that information costs energy. Consider a deck of cards. When you first open the packaging, the deck is ordered numerically and by suit, which is only one way of organizing the cards. Every shuffle creates a unique sequence of cards and takes some energy. There are more possible ways to configure a deck of cards (8×10^{67}) than there are atoms on Earth! Information is embedded in the physical ordering of a shuffled deck of cards; information that can only be revealed by turning each card over in turn—there is no calculation that can work out the order for you. Of course, the information stored in a shuffled deck of cards conveys no intrinsic meaning—unless we assign one—but it is information nonetheless.

Our ability to assign meaning to physical information and store it in matter is the foundation of the global digital economy. Indeed, a biological organism's ability to exploit information stored in matter is of paramount importance in generating excellent fabrication without lavish use of energy.

At its simplest, a circular economy is an economy that can select, from the deck of all possible arrangements of matter on Earth, the arrangement we want now, using the least amount of energy and the smallest number of shuffles, while exporting the entropic cost into space as heat rather than as waste matter into the atmosphere, ocean, or a landfill, which tends to mess up the organization we are trying to achieve. That is to say, an economy that can write the information about our chosen design into the matter of Earth. And our first foray into the profound idea of inserting information into material begins with the idea of additive manufacturing (AM), more commonly known as 3D printing.

The foundations of circularity

In his book, *The Circular Economy: A Wealth of Flows*, Ken Webster describes five principles that characterize a circular economy, which we interpret here in the context of bioinspired additive manufacturing.

The first principle, which we have already discussed, is that material cycles have no beginning and no end. This principle aims to disrupt the unhelpful concept of an "end product" and encourage thinking in terms of flows of matter—such as carbon, water, nitrogen, and so forth—in biogeochemical cycles (page 70). In this view, matter only temporarily resides in a product or organism, as part of a continuous cycle.

The second principle asserts that the size—in terms of geographical area and level of activity—of the flow is of critical significance. The shorter the recycling loop the more profitable and resource-efficient they become, and the easier it is to find mutualistic efficiencies, particularly at the molecular scale. We saw an example of this phenomenon in chapter 3 when we encountered the lossless transport of ATP by diffusion within the cell when compared to the wasteful transmission of electric current over great distances.

As we saw in chapter 1, the rate of material reorganization relates to the amount of energy we apply. The third principle of circularity states that the speed at which matter is reorganized within the loop is crucial. The slower the flow of material, the more efficicient the stock management. This principle resonates with the idea we keep returning to, that sunlight could provide us with a low-intensity energy source to make many small changes spread over time, rather than using high-intensity energy to make fewer, rapid, large changes, and exporting our economic inefficiency as material entropy to landfill.

Insisting on the fourth principle—continued ownership of goods—helps to slow the speed of material flows down. Instead of selling an old product and buying a new one, it could be more cost effective to repair the existing product, keeping the product in the hands of the original owner and saving on energy, materials, and transaction costs. Designing products with reuse, repair, and remanufacturing in mind would facilitate this. The final principle is that our circular economy is still a market economy. We still need to be able to buy and sell materials and goods, and manufacture must be governed by supply and demand.

Ecosystems implement several of these five principles automatically by using universally accessible reservoirs of material—soil, waterways, and air. These reservoirs could be thought of as a marketplace where material exchanges take place, linking the material flows of different organisms and allowing them to find a "coincidence of waste," where one organism's trash is another's treasure. We only need look to dung beetles to understand that utility is in the eye of the beholder. Well-stocked reservoirs enable a rich diversity of different species to occupy the same site, leading to the smallest possible material loops.

From these reservoirs, biological organisms are able to extract the required materials and group them into larger assemblies, such as DNA and proteins. All biological organisms, from dung beetles to dolphins, are masters of such additive processes—they acquire resources and assemble new materials one molecule at a time to drive the endless cycle of growth, repair, and reproduction, while retaining ownership of many of the same atoms, cycling them in the body with only incremental changes.

Directly emulating biological manufacturing in our own circular economy would require us to develop a ubiquitous reservoir of material—analogous to soil— alongside an additive process for combining such a reservoir into useful structures, which is analogous to biological growth. Although this might sound far-fetched, the very early beginnings of such a revolution have been realized in the form of AM, which can be thought of as using a computer program (an algorithm) to combine small amounts of matter into a larger pattern under the direction of an external source of information. If scientists and engineers can master the ability to use information to change the properties of our manufactured materials—like nature does—we may be able to design a colossal variety of extremely sophisticated objects from a finite stream of material, without blowing our entropy quota on energy-intensive processes. A biological cell encodes such manufacturing information in DNA, but it also includes the machinery to interpret and repair that DNA. In an analogy where our homes emulate living matter, each building would be able to extract atoms from a shared reservoir and manipulate them using locally stored digital information.

Under these circumstances a circular economy could move economic incentives away from maximizing the throughput of materials to designing and creating novel ways to insert information into the matter that we already have, using the free and near-limitless supply of solar energy.

This begs the question, "how do you introduce information into material?"

The game-changer of additive manufacturing

During AM, information about the shape of a workpiece is introduced during assembly via a sequence of instructions known as geometry code (G-code). Originally invented to control automated milling machines, G-code works nicely for 3D printers too. Each instruction, such as start, stop, move, print, and so on, guides the printer to add the next morsel of matter to the object, serially introducing information to the manufactured object.

AM allows an incredible range of geometries that are difficult to construct using any other methods. They work by repeating a single operation—"add"—many times to build up an object. Since they only have one operation, 3D printers have simple designs making them cheap for domestic markets and easy for the user to maintain by printing out replacement parts. Indeed, some printers, such as the Rep-Rap, can replace entire printers. Rep-Rap is an open-source desktop-3D printer that was designed to self-replicate (with a little help from humans) by printing all the components needed to rebuild itself.

In many cases, AM can eliminate manufacturing steps such as fabricating molds, masks, templates, or die formers. In theory, entire production lines can switch to new products by changing the information in the G-code, requiring no hardware modifications. This equips companies to respond rapidly to changes in manufacturing demands. In practice, prints often require post-processing steps like rinsing or curing with ultraviolet light or heat, but the overhead cost of changing the production model is still usually much lower than regular manufacturing.

Since 3D printers can be small and easy to maintain, and they can be sited virtually anywhere—right at the point of use or at the site where the input material (the feedstock) is made. Such co-locality eliminates transport steps (see chapter 3) and enables smaller material flow loops to emerge. For example, medical 3D printers produce a wide range of implants in hospitals and surgeries, and if necessary, inside the operating room, which facilitates strategic expert input into their design.

An information-centric manufacturing process allows for the rapid iteration of design and testing, releasing full creative potential and enabling collaboration among designers. In fact, the expression of objects as information means computer algorithms

Opposite, top: A technique called generative design uses computer models to find ultra-efficient structures by simulating designs and optimizing them against arbitrary criteria. The resulting designs are often very organic looking, which is not a coincidence. Living organisms have optimized their designs over billions of years through such an iterative process.

Opposite, bottom: Dutch robotics company MX3D designed a mobile 3D printer that could print its own support and then move along to extend that support, enabling it to install a remarkable and beautiful generativity-designed steel bridge known as the Vondelpark Bridge. Could this be the future of architecture?

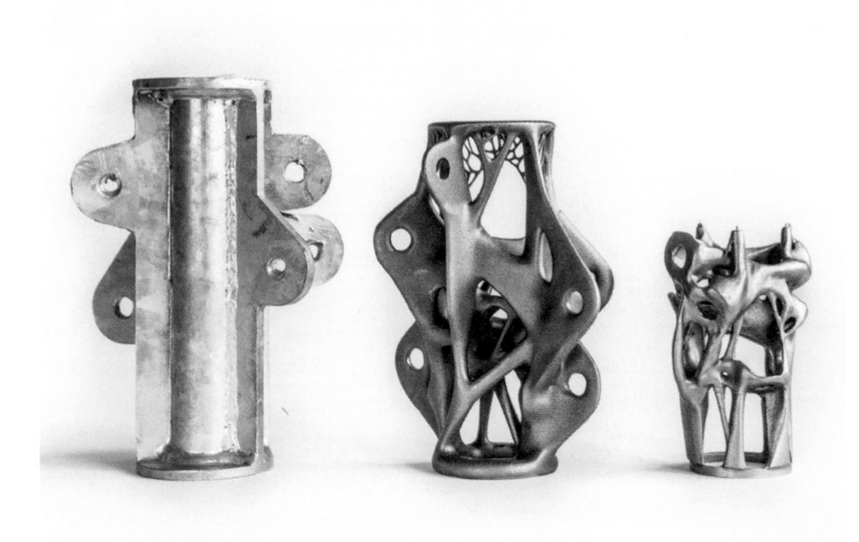

can be used to test the efficiency of different designs before implementing them in material, allowing for complex, highly evolved designs prior to spending energy and material. While biological systems had no choice but to test designs by implementing them in the real world—which took billions of years—encoding our designs as abstract information allows humans to use computers to speed up the selection process and to considerably reduce the resources needed for design.

Additive manufacturing avoids wasting energy and materials in energy-intensive processes—instead small amounts of energy are input throughout the entire serial manufacturing process, meaning that the peak amount of energy used at any one moment (the power) is small compared to manufacturing techniques like smelting or firing. If the maximum power required to print an object is low enough, then it may well be possible to drive the additive manufacturing with sunlight. Such an achievement would dramatically reduce greenhouse gas emissions from manufacturing, and might be just the strategy we need to bring our emissions down below the threshold at which natural photosynthesis can stabilize atmospheric CO_2 within the planetary boundary conditions.

Types of additive manufacturing

With that strategy in mind, let's take a look at the three main classes of 3D printing: stereolithography apparatus (SLA), fused filament fabrication (FFF), and selective laser sintering (SLS).

SLA uses computer-controlled UV lasers or light projection to trigger patterned chemical reactions within a tank of liquid light-sensitive resin. The light solidifies the resin in precise locations. Fresh resin is then introduced and layers are built up one at a time. Chemical additives can give the finished product different mechanical properties such as transparency, heat resistance, toughness, or strength. For example, heat resistant objects that retain strength while warm can be used to support motors that dissipate heat. Alternatively, toughness, strength, and durability may be required for demanding applications such as prosthetic limbs or dental implants. However, the material used is typically a thermoset plastic—meaning the hardening process is *irreversible*—making recycling difficult.

With SLA the level of detail is very high: the most advanced experimental printers can print details as small as tens of nanometers. It is also generally faster than other techniques, since scanning a laser beam is faster than moving a print head.

Printing techniques

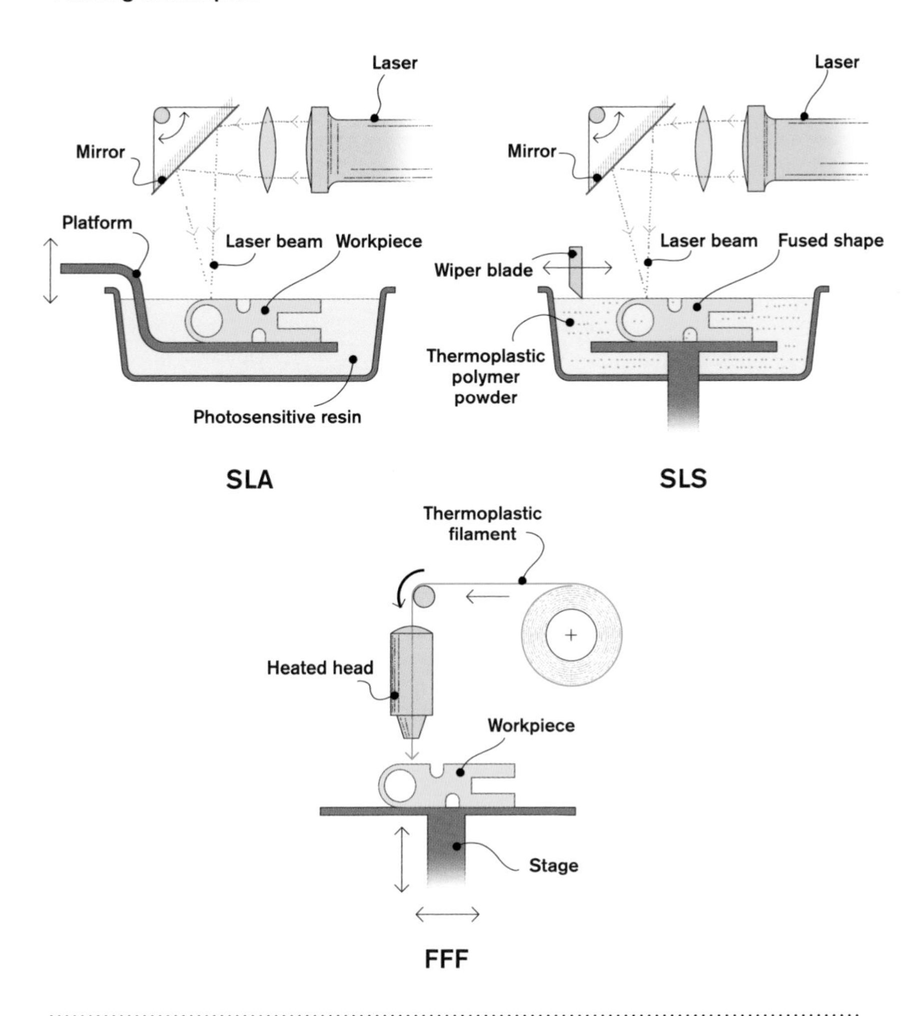

Above: Stereolithography apparatus (SLA) uses a laser or projected UV image to solidify resin. Selective laser sintering (SLS) uses a laser to melt plastic or metal powder into a solid structure. Fused filament fabrication (FFF) uses a spool of material pressed against a hot nozzle to extrude molten material in a manner reminiscent of piping frosting onto a cake.

In FFF—probably the most widespread approach to 3D printing—the print head moves relative to the workpiece. A filament with a low melting point is pressed against a heated plate containing a small hole and then a thin stream of melted material emerges. Softened material is produced just above the workpiece and merges with the previous layer to cool into a continuous solid. The FFF process requires materials that can be melted and reformed, so finished models themselves can be recycled by melting them down, making it easy to close the manufacturing loop.

The final major class of 3D printing—SLS—involves scanning a laser across powdered material. Powder particles fuse together wherever the laser hits, forming a solid structure with a specific design. A new layer of powder is introduced, and the cycle repeated. The un-fused powder acts as a support, enabling a wide range of scaffold-free geometries.

SLS can print metal objects strong enough for engineered systems such as airplanes and rockets. Indeed, Boeing have incorporated such printed objects into commercial and military production units since 2002 and NASA 3D prints rocket cones. But there are challenges. The metal powder grains are randomly oriented making it hard to control the crystalline structure of the finished metal, meaning that one copy of a printed component might differ mechanically from others. However, progress continues to be made in controlling the mechanical properties of metal workpieces produced via SLS.

CLIP™

The key to excellent 3D printing is to control the chemistry and physical conditions in the print zone. Amid an active field of research, one example offers some insight: an additive manufacturing company called Carbon, Inc has developed an intriguing method of 3D printing called CLIP™ (Continuous Liquid Interface Production™).

The technique infuses a thin layer of oxygen at the bottom of the resin tank that inhibits solidification. Solidification only happens when the laser beam hits resin a small distance away from the bottom, where the oxygen level is lower. That way there is always a layer of fresh material present beneath the workpiece, which can be slowly retracted as the object builds up, layer by layer. In other commercial forms of SLA, the workpiece must be alternately raised and lowered to create a fresh layer of resin at the bottom of the tank, which slows the printing process considerably.

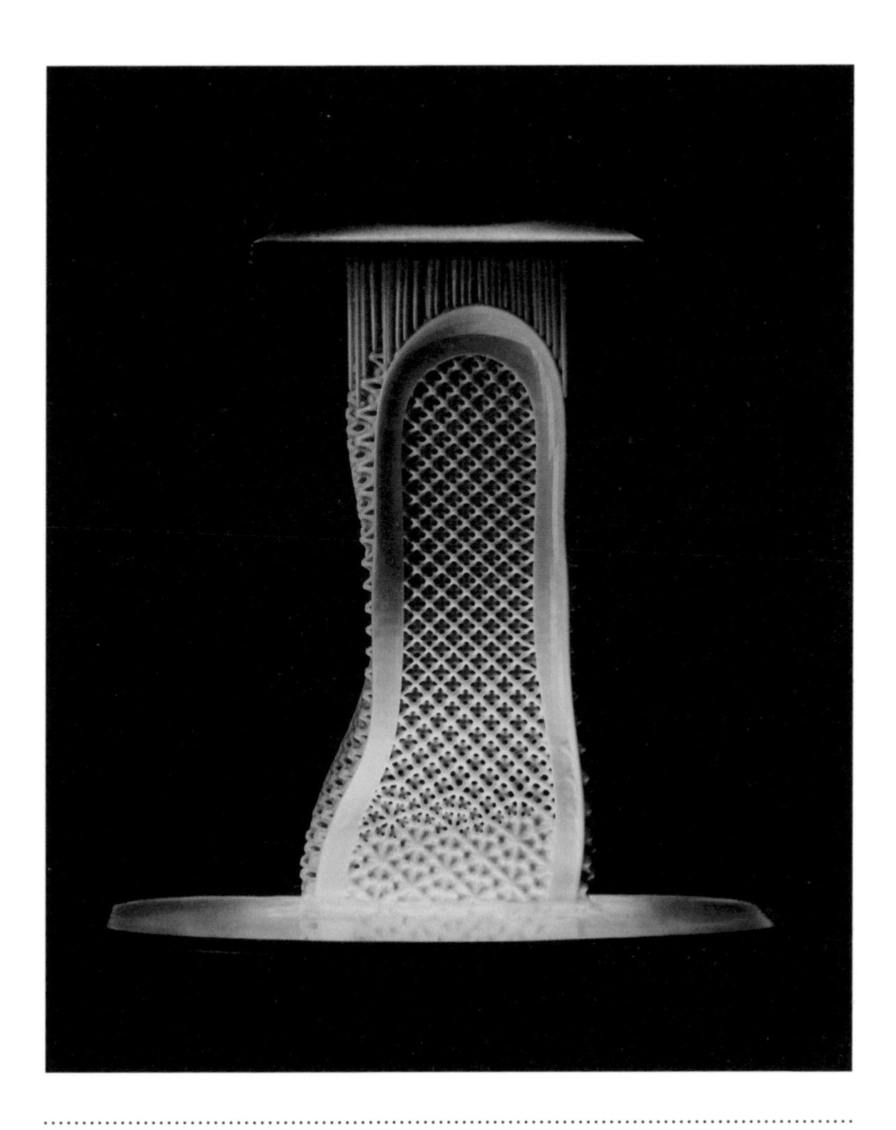

Above: Carbon, Inc's innovative 3D printer uses light to solidify resin and oxygen to inhibit the process. Maintaining a careful balance between light initiation, inhibition of solidification by oxygen, flow rate, and solidification rate using the Continuous Liquid Interface Production™ (CLIP™) system accelerates the printing process between 25 and 100 times compared to other commercial SLA systems.

The drawbacks of 3D printing

However, returning to the goal of creating an AM process that could emulate biological systems, we see that current technology falls short of living organisms' ability to additively manufacture using mixed-material feedstocks. Because each material requires a distinct AM process, printing with mixed materials is difficult, expensive, and makes recycling more complex, but it is not impossible. Furthermore, because 3D-printed objects are printed in layers they tend to be mechanically weak between layers. Nevertheless, improvements are being made all the time and new types of AM are being developed to expand the material range, which already includes plastics, metals, and ceramics—and chocolate, sugar, and gelatin!

While a press can stamp out many thousands of parts per hour, a 3D printer can take many hours for a single print. However, manufacturing speed becomes less crucial in the circular economy because the focus shifts instead to building objects that are more durable. Thus, the drivers for innovation in a circular economy are quality, durability, ease of operation, and maintenance, and low power consumption—challenges that 3D printers are well suited to address.

Nature's 3D printers

To improve quality and durability of 3D-printed products, we need precise chemical control, which is exactly what molecular biology offers. In fact, a remarkable biomolecule called ribosome is, in effect, a molecular 3D printer that biological systems use to build proteins. We can define a 3D printer as a system that needs a source of energy to combine input materials into a workpiece under the control of a sequence of information. In a similar vein, the ribosome takes ATP (energy) to join together chains of amino acids (the input materials) to form a protein (the workpiece), controlled by the DNA sequence (the information).

While the instructions for 3D printers are written as binary numbers—ones and zeros—the instructions for a ribosome are written in an alphabet of four molecules called adenine (A), cytosine (C), guanine (G), and thymine (T), which are examples of a kind of molecule called a nucleotide. We have met adenine already—it is the A of ATP. Indeed, adenine is just one of many building blocks for which biological organisms have found multiple uses.

The language of life

To encode instructions for building a protein, the four molecules, ATCG—known collectively as bases—are each attached to a sugar molecule called ribose that is stripped of an oxygen and assembled alongside other ribose-base combinations. The result is a long strand of deoxyribonucleic acid—to give DNA its full tongue-twisting name. Two of these strands twist together to form the famous DNA helix whose sequence of nucleotides encodes blueprints for all life on Earth.

Within the DNA code, nucleotides group into triplets, known as codons. If nucleotides are the letters of DNA, codons are the words. The dictionary of life contains codons for the 20 amino acids used to build proteins, as well as other useful words, like "stop" and "start." The codons are the equivanlent of the G-code controlling 3D printers.

To prime the biological additive manufacturing process, the DNA sequence is copied into a molecule of ribonucleic acid (RNA)—DNA's single-stranded cousin. The RNA's job is to carry the information from the nucleus, where DNA is stored, to the site of protein construction, and so it is known as messenger RNA, or mRNA for short. The mRNA strand threads through the structure of the ribosome and as it moves through, the encoded protein chain is constructed—one amino acid at a time.

However, there is a major difference between ribosomes and 3D printers: the ribosome works with a multicomponent feedstock—20 different kinds of amino acids floating around in the gel-like cytoplasm of the cell. This would be equivalent to a 3D printer that could print solid red or blue objects from a purple resin that was a mix of red and blue pigments. The ribosome can distinguish between different amino acids because each one is attached to a molecule called transfer RNA (tRNA), which will add its amino acid to the growing chain only if it matches the codon sequence of the mRNA.

Circularity built-in

The principles of the circular economy described by Webster are inherent in this protein-making scheme. Small loops within a single cell maximize material and energy efficiency. Molecular 3D printing produces proteins only where they are needed, such as the enzymes that help to harvest energy, which are produced directly in mitochondria. This energy, in the form of ATP, then powers the ribosome 3D printer to fabricate more proteins.

There are also specialized proteins, called proteases, whose task is to chop unneeded proteins up into amino acids again (akin to recycling), and enzymes called tRNA synthetases, which restock the amino acids on tRNA molecules (akin to refilling the feedstock). The whole system includes loops of many different components, all interlocked within a very small volume—a single cell.

Every single cell in your body is steadily manufacturing new proteins all the time. The effect of such parallel, rather than centralized, manufacturing is dramatic. For example, although it takes about one minute for a ribosome to make a protein with 200 amino acids, each cell contains around 10 million ribosomes and there are around 37 trillion cells in your body. Altogether, your body can produce 22–44lbs (10– 20kg) of protein every day. That's a lot of material to produce with atomic precision, and it uses up around a fifth of your daily energy (equivalent to the energy contained in 1.6 sticks of dynamite).

By co-evolving compatible sets of enzymes, organisms come together to form complex networks that control the flow of materials through the living and non-living components of the ecosystem. When we cut down forests and replace them with buildings made from concrete or steel, or when we grow vast single-species fields of crops, we are literally destroying the ability of whole ecosystems to process local chemicals and fabricate a wide array of materials and structures using direct sunlight as the energy source—a spectacular own-goal!

The ability to produce proteins via an information-controlled additive process allows living organisms to perform atomic-precision manufacturing and chemistry over an area of land spreading out over thousands of miles, and all powered by sunlight. We should do the same.

Harnessing molecular manufacturing

Perhaps our energy-intensive industrial processes could be entirely replaced with more intelligent information-based local processes that could be powered by sunlight. For example, one study classified different biological and technological solutions to various problems according to the categories of: energy, information, time, space, substance, and structure. Humans favor high-energy solutions whereas organisms make much more use of information, time, and structure—particularly at the scale of tens of micrometers, which is about the vertical resolution of a good desktop 3D printer. Indeed, biological material stores information at least a million times more densely than the best current technology. A typical hard drive for a modern laptop measuring 0.9 x 1.65 x 0.06 inches

Ribosome

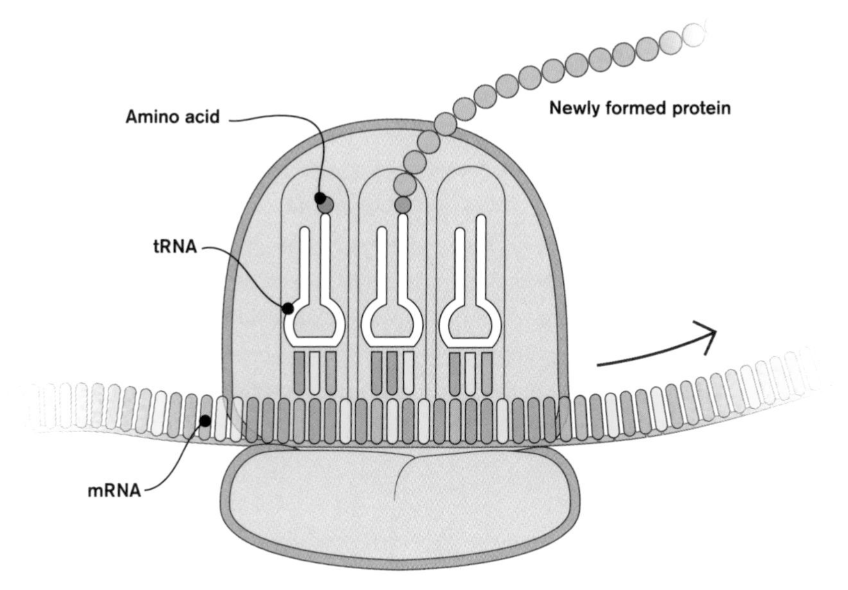

G Glycine
P Proline
A Alanine
V Valine
L Leucine
I Isoleucine
M Methionine
C Cysteine
F Phenylalanine
Y Tyrosine
W Tryptophan
H Histidine
K Lysine
R Arginine
Q Glutamine
N Asparagine
E Glutamic Acid
D Aspartic Acid
S Serine
T Threonine

This page: Proteins are constructed using molecules called tRNAs, which each carry one of the 20 amino acids used by living organisms. Each tRNA molecule has a molecular barcode known as a codon, which corresponds to a particular amino acid in the genetic code, shown in the wheel (right). An mRNA molecule carrying the sequence to be translated winds around the ribosome and acts like a barcode scanner. Only the correct tRNA—carrying the desired amino acid—can dock with the ribosome and add its amino acid to the growing protein chain. The genetic code is made up of four nucleotides—A, T, C, and G—which form the barcode sequences for each codon (reading outward from the centre of the wheel).

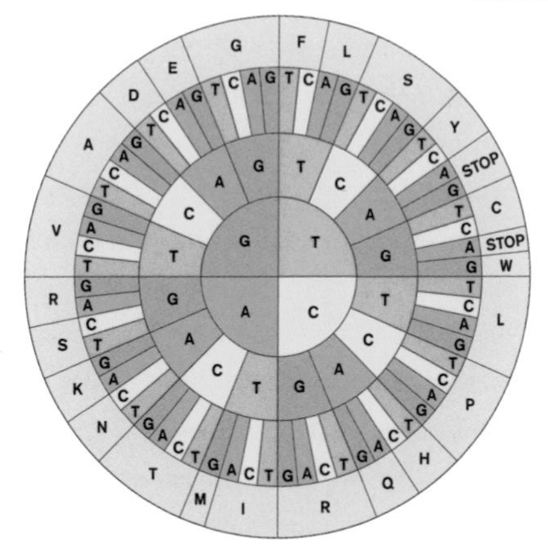

(22mm x 42mm x 1.5mm) has about the same volume as 411 million eukaryotic cells. While the hard drive might store 2 terabytes (TB), the same volume of cells could store a million times that amount of information (1.4 exabytes; EB).

Mastering molecular-scale information storage and algorithmic manipulation of that information might allow us to use ingenuity found in the biological loop to upgrade the processes in the technological loop of the circular economy. Perhaps our manufacturing systems could rival biological ones in terms of sophistication, but using a set of completely different base materials? A spectrum of technologies could be designed—from single-component feedstocks supplying serial 3D printing, to multicomponent feedstocks supplying parallel printing processes—that operate all over the world. The smaller and closer to one another those printing processes were, the more efficient the circular material loops would be, and the more likely that beneficial synergies between distinct processes would become apparent.

Dynamic additive manufacturing

All the examples of AM so far, except for protein, have been static objects. What if we could introduce an element of dynamism? The most advanced 3D printers allow users to manipulate the parameters of the printer, such as temperature, flow rate or nozzle distance and so on, *during* the printing process. These innovations complicate the printing process but introduce much more flexibility, which will lead to novel applications. For example, strong yet flexible materials whose density gradients vary, such as bamboo, which is used routinely as a construction material for houses, boats, and scaffolds for sky scrapers, has also been used to construct turbine blades.

Such time- or position-varying conditions are routinely exploited by biological organisms during the growth process. For instance, the pigments in your iris that produce browner shades of eye color only develop in reaction to light, meaning all babies' eyes darken in the first few days or weeks after birth. Another example is spider silk, an impressive material about the same tensile strength as steel, which is manufactured under a gradient of phosphate ions in the duct leading to the spinnerets on the spider's abdomen. The gradient modifies the interactions between the proteins, forcing the structure of the silk protein to change as it is drawn through the duct.

Another branch of materials science is looking at is 3D-printed structures that change dynamically in response to environmental conditions. So-called 4D printing is

Energy vs information in human and biological systems

Engineering solutions

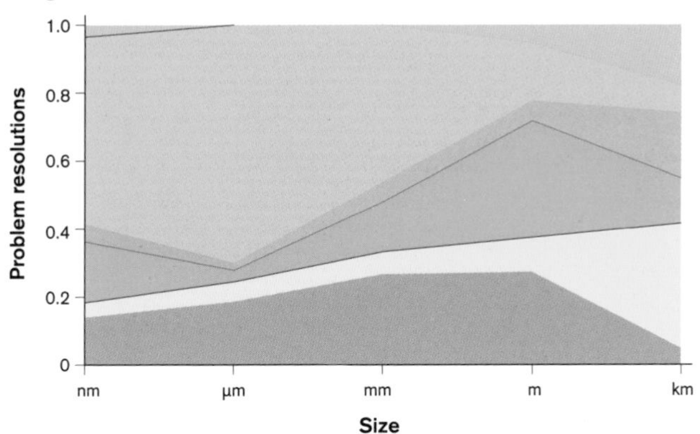

Biological solutions

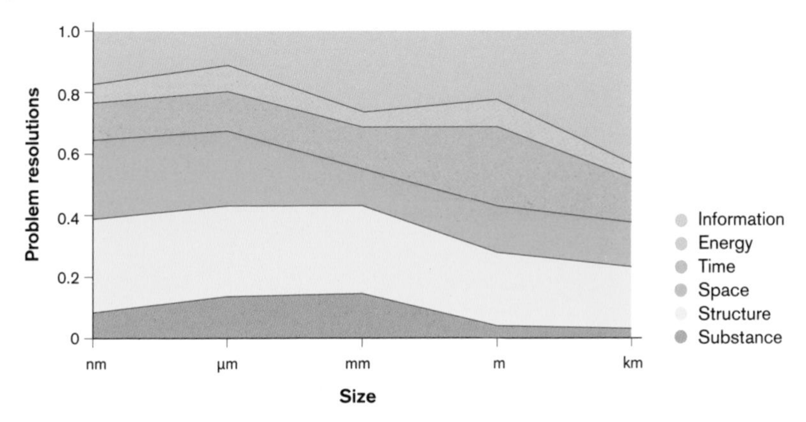

This page: Humans overwhelmingly rely on energy-based solutions to rearrange matter—particularly at smaller scales—which contrasts heavily with the tendency for biological organisms to avoid energy expenditure. Biological systems most commonly employ structure- and information-based solutions, particularly at larger scales. These observations were made by scientists by categorizing solutions to a series of engineering and biology challenges according to their use of energy, information, time, space, substance, and structure. Here, the solutions found in each category are distributed visually across different scales for the two domains.

already helping engineers design entirely new classes of robot made from soft materials that change shape. A great example would be a gripping device that changes shape dynamically in response to the object it picks up, such as the jellyfish grabbing robot designed by Nina Sinatra and colleagues at Harvard. 4D-printed objects need a power source to change shape, so depending on the nature of the object, that energy could come from the environment in the form of heat, light, hydrodynamic or pneumatic forces, or from internal potential energy generated during construction, resulting in a "once only" movement that could be useful for self-assembly or self-recycling.

Dissipative systems

4D objects exchange energy with their environment to drive a change in their structure, but they don't exchange matter. Objects that exchange both material *and* energy with their environment—such as biological organisms—are called dissipative systems. Specifically, a dissipative system is open with respect to both energy and matter, and is far from a state of equilibrium. Such a description applies to a huge range of scenarios including biological cells and organisms, tornadoes, rivers, ecosystems, mountains, stars, galaxies, and pretty much every collection of objects—animate or inanimate—in the known universe.

A system that is in equilibrium has no net exchange of matter or energy with its environment and anything that you can measure about the system, like its volume, pressure, or temperature, is stable over an *infinite* amount of time. A leak-free balloon that has been inflated and tightly knotted could be said to be in equilibrium. Dissipative systems are the opposite—they are constantly changing—and this could make such systems intrinsically easy to recycle. For example, since a dissipative system is reliant on a continued energy supply, cutting that off allows the system to collapse back to its constituent parts in order to reach equilibrium, just as when an organism dies its materials degrade and are returned to the reservoirs of soil and air from which they came.

The branch of science that deals with such phenomena is called non-equilibrium physics and even though the field is well over a hundred years old, the science is still not fully understood. However, a great deal of theoretical progress has been made in recent years, building on the pioneering work of Ilya Prigogine, who won the Nobel Prize in Chemistry in 1977 for proving that order could spontaneously arise from highly dynamic dissipative systems that were far from equilibrium. Prior to this, in 1968, together with his colleague René Lefever, Prigogine presented a useful theoretical

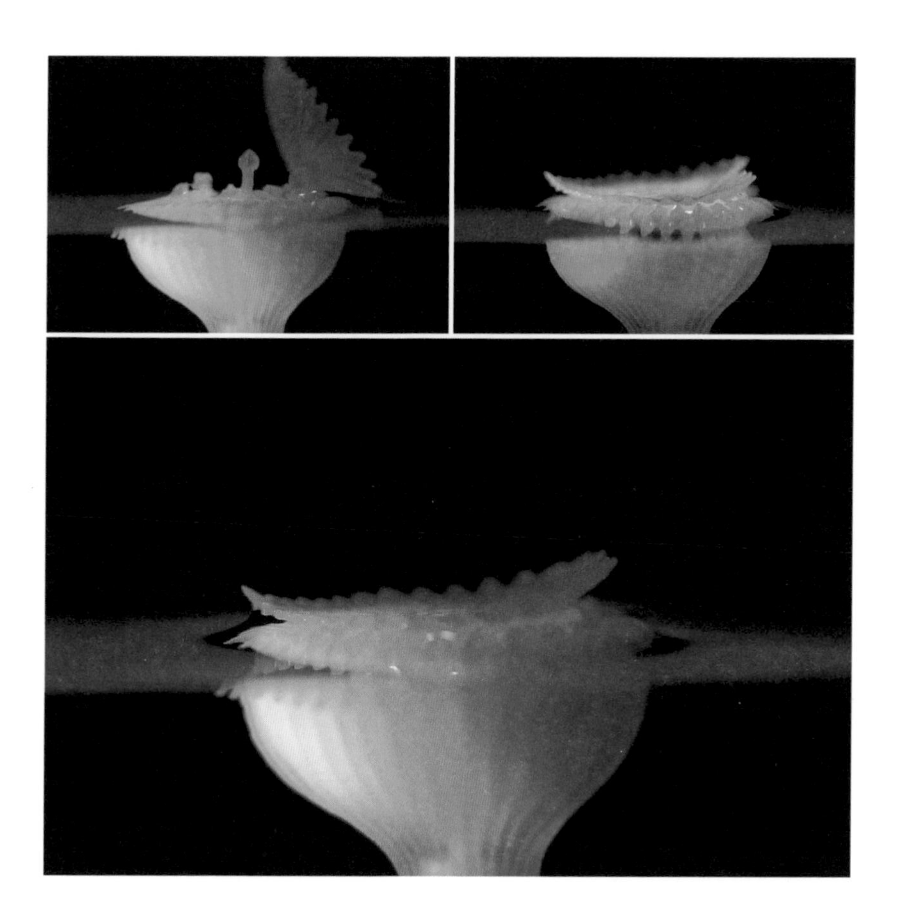

Above: Multi-material 3D printing allows for creative designs that are dynamic. These clever 4D designs are driven between two states by a pneumatic pump and were created by artist Nicole Hone using a Stratasys Connex 350 3D printer.

model—later called the Brussellator—which explains how chemicals can react in complicated networks of reactions causing them to exhibit oscillatory behavior.

Earth itself is in a certain type of non-equilibrium arrangement called a steady state. In such systems, material or energy enters and leaves the system in roughy equal amounts—like turning on the tap in the bath and pulling out the plug. Work is done continuously on the constituent parts, which are rearranged endlessly, yielding a constant stream of entropy out to the universe. On Earth, solar energy enters and leaves the system very rapidly, whereas matter is gained and lost much more slowly.

A system need not be changing *quickly* to be a dissipative system. The Himalayas formed when India collided with Asia about 40 to 50 million years ago, and the tectonic plates in the Earth's crust are still settling down. Another example is a glacier, which tends to change so slowly it seems almost static. However, glaciers also show us that a dissipative system can change pace if the amount of energy in the system changes—some glaciers in Greenland are now moving several meters per day as a result of a warmer climate.

Our theoretical understanding of non-equilibrium systems is not complete, but experimentalists, simulators, and theorists have succeeded in exploring systems that generate dynamic structure. Such matter is termed active matter. For example, large groups of molecules in little sacks can be moved around the cell by ATP-fueled motor proteins that "walk" along a rigid protein microtubule known as the cytoskeleton. These cellular components can be extracted to create artificial dissipative systems.

The need for synthetic biology

An entirely untapped field of research awaits, in which far-from-equilibrium phenomena yield steady state systems that could collapse into equilibrium with their environment at the end of product life. In this case, matter would only be on loan from the environment, with energy applied to drive the matter into a useful dissipative state and keep it there. In developing such a system, we would open up a new approach to product lifecycles.

Biological systems already integrate dissipative processes, information-driven control over chemical reactions, and adaptive responses to the environment. Consequently, synthetic biology—a discipline devoted to designing and manufacturing novel biological components and organisms—also has full access to such advanced concepts. It is no coincidence that the industry, built on the back of 50 years of molecular biology research—which gives us the power to reprogram cellular DNA to produce virtually any compounds that we conceive—has had over $18 billion investment from private and public sources worldwide.

What exciting technologies based on the concepts of 4D printing, dissipative systems, and synthetic biology will emerge over the next 50 years? Just as biological cells harvest energy and manufacture products locally, such technology may well be applied to our homes and cities, with the smallest possible recycling loops. Zooming out to regional and global scales, such technology could simplify circular economics, while reconnecting us with nature and teaching us to create non-biological systems that sit harmlessly alongside a free and wild biological world—as rich as it was before humans.

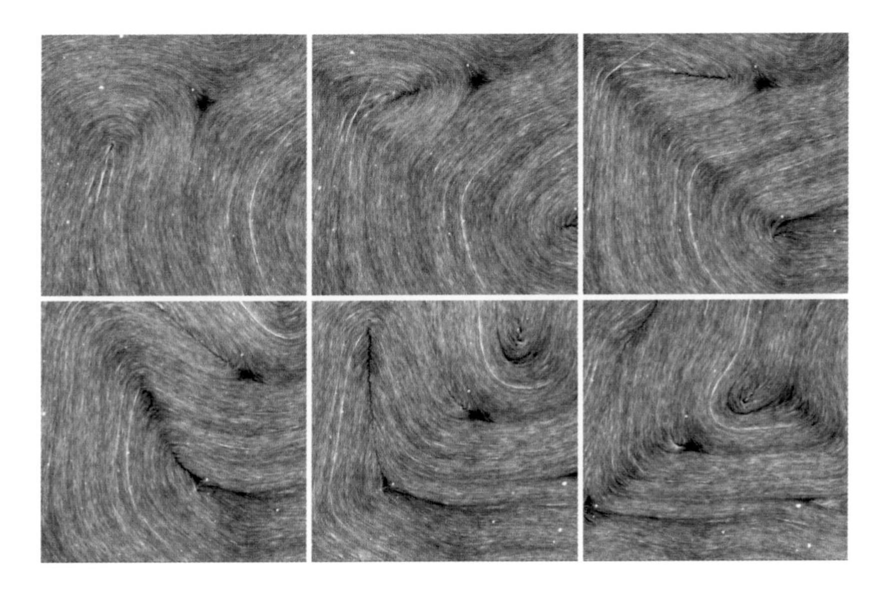

Above: Scientists Linnea Lemma and Zvonimir Dogic were able to make these timelapse images by extracting the microtubule system from a biological cell. Bundles of microtubules are moved relative to each other by millions of kinesin motor proteins, forming large-scale swirling structures that resemble fingerprints. The defects seem to take on a life of their own, moving through the structure. Perhaps in the future we will be able to control such structures in dynamic printing methods, and freeze them in place, just as such patterns are frozen on your fingertips.

Printing electronics

In the future, a fully working smartphone could well be 3D printed one atomic layer at a time. However, such an approach isn't practical—the smaller the details we want to print, the slower and less accurate printing becomes—although a few optimistic engineers are trying. Printing an entire smartphone this way requires printing 30 chemical elements with atomic precision; the slightest misalignment and the components wouldn't work. Unfortunately, atoms are never still—even atoms in solids constantly jiggle with Brownian motion. The most precise 3D printers—using 2-photon lithography—solidify 100 nanometers (nm) regions of resin meaning that, currently, the smallest printable unit is a 3D blob of matter containing around a billion atoms, far larger than 5nm transistors.

Instead of reaching for atomic-resolution printing, maybe we could assemble materials using conventional chemistry? Living systems use enzymes to build copies of molecular building blocks, such as the amino acids and tRNAs they attach to. Such molecules diffuse randomly, until they encounter ribosomes, which assemble the proteins. Such proteins can be anywhere between 2nm to 25nm across, giving us a potential mechanism for making atomically precise units containing a million atoms. Indeed, researchers have succeeded in making proteins behave like complete transistors.

The same Brownian motion that prohibits atomic printing drives assembly of macromolecules into larger structures, such as conducting fibers or complex electronic systems like the respiratory- and photosynthetic-complexes. Furthermore, just as water and oil separate into droplets, proteins can spontaneously separate based on sequence. Such phase separation could sort millions of proteins into droplets each containing particular proteins. With such a phenomenon, maybe we can generate droplets several micrometers across that contain millions of organized, self-assembling molecules, each one capable of acting as a transistor.

Atomic printing is off the cards, but it is entirely feasible to 3D print structures the size of biological cells. Indeed, researchers have succeeded in printing entire hearts using living cells. Maybe, instead of living cells, we could generate complex and sophisticated polymer inks that self-assembled into organized droplets? We could then print out such droplets in arrays to create bulk systems that spontaneously self-assemble at the molecular scale to form larger objects. 4D manufacturing also opens new avenues; droplet arrays have been made that behave like electronic circuits but can fold into new shapes after printing. Or we may not have to rely on a moving print head to deposit our droplet inks. Dissipative systems, which depend on a driving force, such as light or stirring, could yield complex shapes that we can control.

Combining the power of ribosomal or polymer molecular assembly with advanced 3D- and 4D-printing, spontaneous droplet formation and dissipative systems could be a viable route to the atomic-resolution fabrication of a handheld device in which all the components were assembled in one place.

Biological additive manufacturing

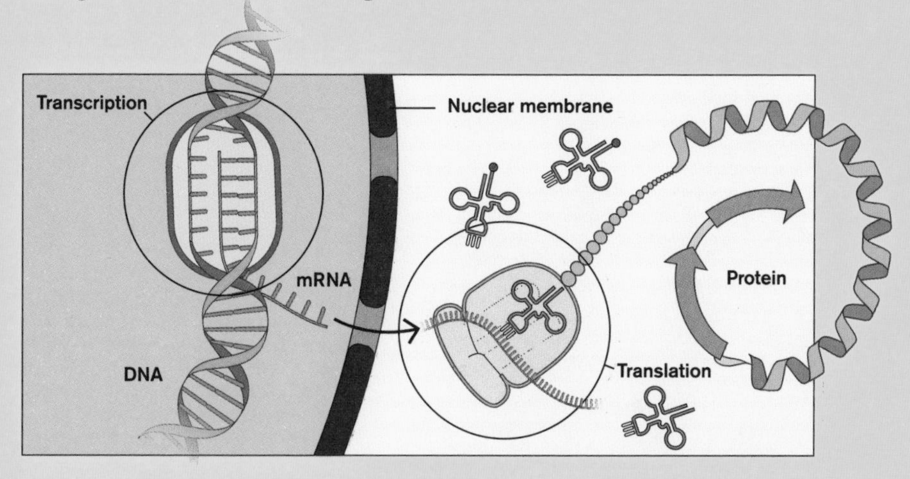

Above: The biological process of making a protein—known as translation—is analogous to 3D printing. A molecule—called a ribosome—plays the role of the 3D printer and employs energy to combine a supply of material (amino acids) into a workpiece (the protein) using a sequence of instructions. The information about the design of the protein is transcribed from long-term storage in DNA into a molecule called mRNA, which the ribosome is able to read. The process requires energy supplied by a molecular fuel, ATP.

5 Synthetic Biology

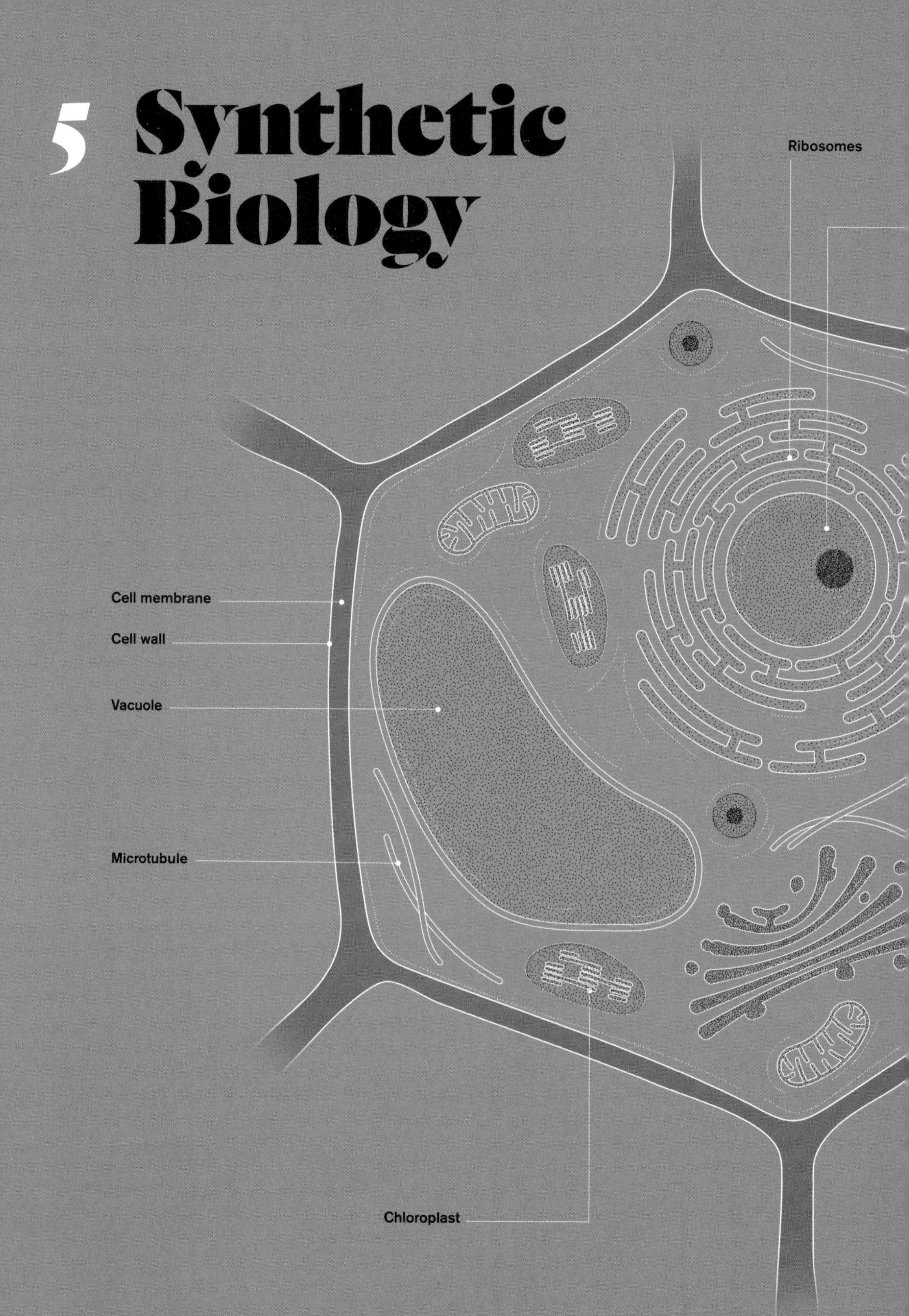

Nucleus

"In the same way that we simplified and abstracted components from physics to allow us to build billion component processors, we can and will modularize, abstract, and understand biological components with the explicit goal of constructing artificial biochemical and biological systems."

Tom Knight, co-founder of synthetic biology company, Ginkgo Bioworks

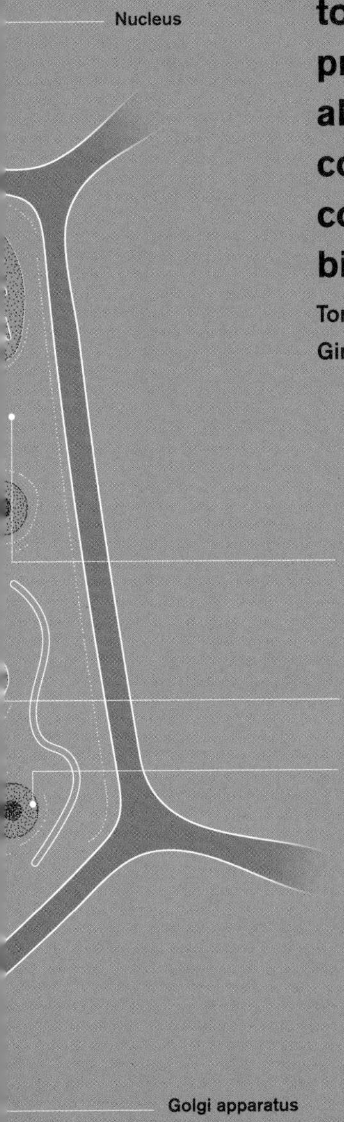

Cytoplasm

Mitochondrion

Peroxisome

Golgi apparatus

Plant cell structure

So far we have learned that biological systems have mastered the ability to read and write information into a whole range of molecules, particularly DNA and proteins, which opens the door to biological additive manufacturing and excellent control over environmental chemistry with the use of enzymes. We may have discovered 3D printing only late in the last century, but evolution has had billions of years to master a whole range of interacting biomolecules as tools to manufacture some incredible materials and functional systems, one molecule at a time.

The question is: can we take this biological process and learn how to use it to balance the impact of humanity on the natural world, while still achieving the quality of life that we now know is possible? Perhaps, if we can understand more about the complexities of biological functions, we can expand our activities into the biological loop of the circular economy while reducing the volume of flows in the technical loop. In doing so, biological systems could provide us with so much more than they already do, but in a way that is fundamentally compatible with maintaining the natural capital of planet Earth.

Storing information in matter

Tightly coiled inside the nucleus of every cell, reams of DNA encode the amino acid sequences for every protein the organism requires to function—from enzymes to signaling molecules.

To put this into perspective, the human genome is about 3.2 billion letters long which—stretched out—is about 6ft (2m). That means that each human *body* contains around 125 billion miles (200 billion km) of DNA; enough to wrap around the Earth about 5 million times. Chapter 4 described how the instructions stored in DNA are turned into proteins by the ribosome using an additive process with two steps—transcription and translation. By learning more about how biological systems use this setup to adapt their functionality over time, we gain insight into how we might adapt it to improve our own technology.

When errors are useful

DNA is formed from a double helix—two chains of molecules spiralling endlessly around each other in an elegant molecular dance, bridged by chemical bonds between the base pairs. But the data doesn't have to be stored in a double helix; single strands of nucleotides can store information just as well.

The second strand in a DNA molecule acts as a backup. The way nucleotides pair up means the second strand is always complementary to the first and therefore carries a mirror image of the original sequence, which can be used to check for mistakes. When DNA is replicated during cell division, the two strands open up, each acting as a template for a new copy. Specialized error-detection proteins can check the newly formed strand against its copy, detecting (and destroying) many new accidental mutations before they have an effect.

Indeed, we often talk about mutations as the fuel of evolution, but most organisms have many sophisticated cellular systems in place to prevent mutations from being passed on. That's because the vast majority of mutations would be harmful. Imagine randomly changing letters in a copy of *War and Peace*. What is the likelihood that your randomly modified words are real words? And how many of the sentences would make grammatical sense?

When a nucleotide mutates, the gene containing that altered nucleotide sequence may no longer work properly. Since most genes perform useful functions, it's safe to assume that most of the time mutations won't be a good thing; many diseases have their roots in mutations that render important genes unable to function. Mutation is a rare event, in the sense that any given nucleotide is unlikely to undergo one in a particular cell division, but because of the sheer size of genomes and the number of cell divisions, mutations happen all the time.

Very occasionally, a change to the nucleotide sequence of a gene in the reproductive cells can result in improved or new functionality. Although vanishingly rare, these tiny copying errors occur frequently enough to generate the variation that has enabled billions of different life forms to evolve and adapt to their surroundings.

Sometimes major copying errors occur that allow a sudden step change in the rate of evolution. Large chunks of DNA, even whole chromosomes, can be moved or duplicated during the DNA replication process. Gene duplications, for example, have been important for organisms to evolve suites of related proteins and enzymes, and whole genome duplications have been a key feature in the evolution of cultivated plants. Over the last ten thousand years, farmers unwittingly selected for genome duplications in their crops because they resulted in bigger plants, more vibrant flowers, and larger fruits, without disrupting the rest of the biochemical processes that keep the plants alive.

The genetic engineer's toolkit

The capacity of DNA to mutate and evolve has allowed evolution to refine biochemistry to create remarkably complex and responsive chains of cellular activity, a fact humans have been exploiting for thousands of years. Initially, we put the right naturally occurring organisms, like yeasts or fungi, into the right conditions with the right ingredients and left them to perform the chemistry. After a few chance discoveries, humans developed the techniques to produce wine, beer, bread, cheese, and many other products. As our genetic technologies developed, we started to take more control over the biochemistry we outsource to microbes.

Our earliest genetic modification technique, developed long before we had any concept of a gene or DNA, was selective breeding. We bred from plants and animals that possessed the traits we wanted to exploit thereby shaping the evolution of domestic species, over thousands of years, into the diversity we see today. Dogs, for instance, come in an almost unending variety of different shapes, sizes, and coat types, but all domestic dogs (*Canis lupus familiaris*) are descended from a single ancestor species—a now-extinct wolf (*Canis cf. lupus*)—and, in theory at least, all breeds can mate to produce viable offspring (the definition of a single species).

In the 20th century, with new understanding of how genes control biological processes, scientists began to modify organisms more deliberately, copying genes from one organism to another. The basic genetic engineering toolkit is formed from six foundational technologies that enable scientists to perform functions we are all familiar with: they read, write, copy, cut, paste, and insert DNA into the genomes of plants, animals, and microbes.

Reading DNA

The ability to read sequences of protein, DNA, and RNA—the first technology needed to understand the intricacies of biological additive manufacturing—began in the 1950s with the work of British biochemist, Frederick Sanger, who won two Nobel prizes for developing protein and DNA sequencing techniques. Since then, genetic technology has advanced astonishingly quickly. The human genome project was completed in 2003 and now more than 3,500 species have had their entire genome sequenced. In terms of speed, genome sequencing technology has increased a million-fold while the cost has been reduced a thousand-fold over the space of just two decades. We can also sequence the entire catalog of mRNAs (the transcriptome) or proteins (the proteome) in an organism to understand which genes are being actively translated and transcribed in the molecular manufacturing production line.

Copying DNA

The second foundational technology was the polymerase chain reaction (PCR), a laboratory technique that copies a specific strand of DNA many times, using heat to break apart the double-helix and create single-stranded templates to copy. This procedure was revolutionized in 1976 with the discovery of an enzyme called Taq polymerase, isolated from a heat-loving bacterium (*Thermus aquaticus*) found in the waters of thermal springs. The microbe's enzymes function at temperatures far higher than those used to break apart DNA strands during PCR, so Taq isn't harmed by the heating stages of PCR and can be used again and again without the need to refresh the enzymes. With automated PCR machines to do the heating and cooling, millions of copies of a single strand of DNA can be made in a matter of hours. The only catch—you need to attach a short sequence of DNA known as a primer to the end of the strand you want to copy so that Taq can identify it.

Above, top left & right: The humble cabbage that we all know and (some of us) love, *Brassica oleracea*, looks nothing like its wild relatives because by selectively breeding for a bigger, faster-growing crop, humans inadvertently duplicated the plant's entire genome.

Above: During genome sequencing, loose nucleotides labeled with different colored fluorescent markers are used in a PCR reaction to create a complementary copy of the DNA being sequenced. The colors can then be "read" by a machine and translated back into the nucleotide sequence of As, Ts, Cs, and Gs. Next-generation sequencing methods can sequence whole genomes rapidly.

PCR revolutionized the study of genetics and had many useful applications of its own—for example, modern forensic techniques rely on PCR to amplify minuscule samples of DNA found at crime scenes into quantities large enough for genetic fingerprinting, which may be crucial for securing the perpetrator's conviction.

Cutting and pasting DNA

A third important development was the discovery of molecular "scissors," known as restriction enzymes, which snip DNA into fragments every time they encounter a precise sequence of letters. These enzymes form part of the bacterial defense against invading pathogens and parasites. A fourth tool in the box is an enzyme called DNA ligase, which can stitch DNA together, allowing scientists to add sections of DNA snipped by restriction enzymes into any other section of DNA.

Writing DNA

We can also write short strands of DNA using a process called artificial gene synthesis, or DNA printing. This technique assembles custom strands of DNA up to 200 nucleotides long without the use of a template strand, making it quite distinct from how biological organisms fabricate DNA. Known as oligonucleotides, these short strands can then be assembled into longer sequences using DNA ligase. This two-step process has been used to synthesize modified versions of naturally occurring genes, chromosomes, and genomes.

Writing genes from scratch, rather than editing or copying from other living organisms, represents a much more significant challenge. Although protein structure is ultimately determined by an amino acid sequence, working out the shape of a protein from a novel sequence of amino acids is an extraordinarily complex calculation; there are more possible protein sequences than there are atoms in the universe, but only a tiny fraction will form useful 3D shapes. So although possible, creating genes from scratch remains a rarity in synthetic biology. As we will see in Chapter 7, super-computing and artificial intelligence can help to make this gargantuan task more manageable and could allow us to routinely program our own genes for novel proteins.

Inserting DNA

The next step in genetic engineering is usually to insert the DNA into a biological cell, which often requires a plasmid—the final tool in the kit—to carry the DNA into the cell.

Above: Proteins such as Green Fluorescent Protein (GFP) taken from jellyfish (*Aequorea victoria*, top left) have proved an invaluable addition to the genetic engineering toolbox, allowing scientists to visually track which cells have successfully taken up the new DNA. GFP is also used in basic research to visualize and track particular cells or cell types. For example, mice (*Mus musculus)* engineered to express GFP (bottom right) have been used in cancer research, and neuroscientists are studying genetically engineered "brainbow" mice (bottom left), whose neurons express different combinations of three modified GFP proteins that fluoresce in different colors. Non-visual labels are usually used for screening engineered plants and animals for commercial applications, except in cases where fluorescence is the desired trait, such as novelty pets like GloFish, the glowing zebrafish (*Danio rerio*, top right).

Plasmids are tiny, circular pieces of DNA found in bacteria that replicate themselves independently of the main genome. Plasmids help bacteria to evolve extremely quickly because they make it easy for cells to "swap" genes. Genes conferring antibiotic resistance, for example, are often passed horizontally from bacterium to bacterium on plasmids. By the same virtue, plasmids are a powerful tool for genetic engineers wanting to get new DNA into cells.

For example, the bacterium *Agrobacterium tumefaciens* infects plant cells by injecting plasmids into the cell, which then force the plant to divert resources toward producing more *Agrobacterium*. Scientists discovered that by removing the infectious genes on the plasmid and replacing them with DNA we want to insert into a plant cell, they could harness the bacterium's natural ability to modify plant genomes.

These trial-and-error approaches do yield results, but they require many hundreds or thousands of attempts for just one success. For this reason, methods to screen modified plant and bacteria cells are essential for genetic engineering. Commonly, scientists do this by "labeling" their inserted DNA with another gene that they can select for at a later stage. Examples of such labels include genes that make mutant cells resistant to certain drugs or cause the cells to produce fluorescent proteins.

Precision editing

None of the methods explored so far allow scientists much control over where DNA is inserted. The precision needed to insert genes in specific locations only arose with the development of a remarkable genome-editing system known as CRISPR—another technique stolen from bacteria. The CRISPR–Cas system was discovered in 1993 by Spanish microbiologist Francisco Mojica, whose research into the genome of a salt-loving microbe (*Haloferax mediterranei*) revealed a strange pattern—the genome contained the same short sequence, repeated many times back to back. Sandwiched between these repeats were short chunks of DNA that looked suspiciously like viral genes, leading Mojica to hypothesize that these sequences formed part of a defense system against viral infections.

The discovery allowed scientists to develop a precise system for targeting and editing DNA, known as CRISPR-Cas9, that can be programmed digitally to alter the genome and introduce a specific DNA sequence at a specific site. By attaching a guide sequence to a DNA molecule, researchers can direct CRISPR-Cas9 to store it in the bacterial

CRISPR

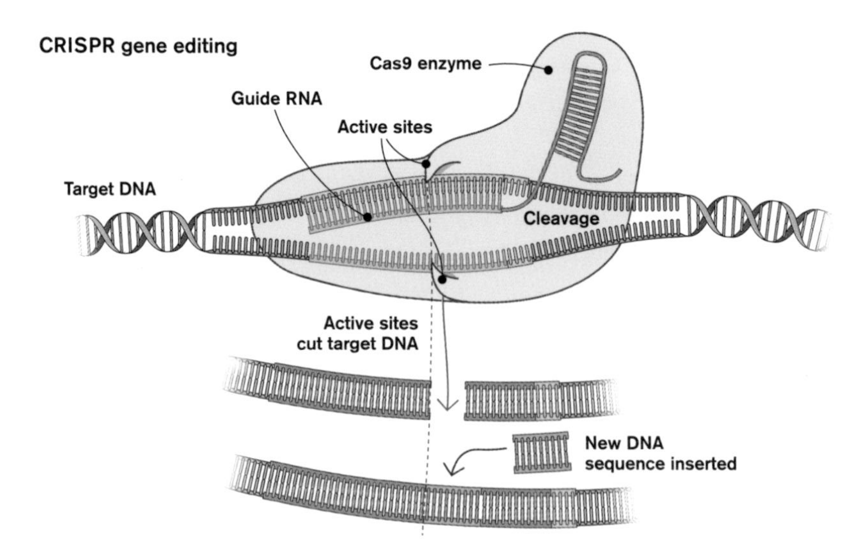

Natural bacterial immunity

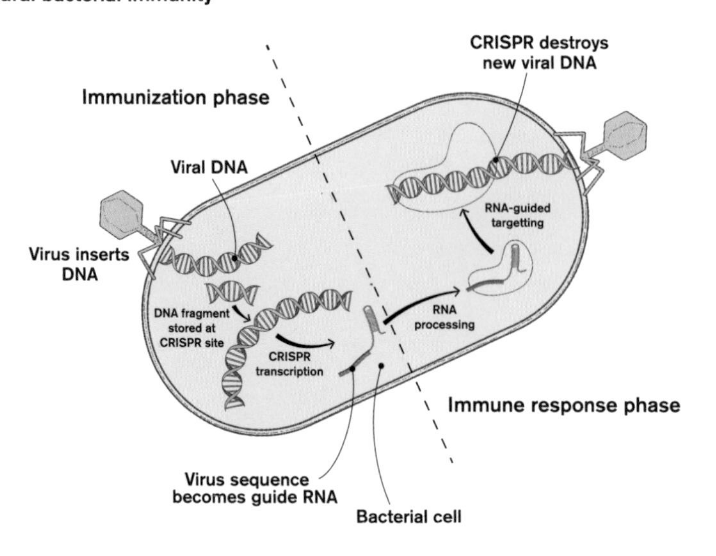

Synthetic nucleotides

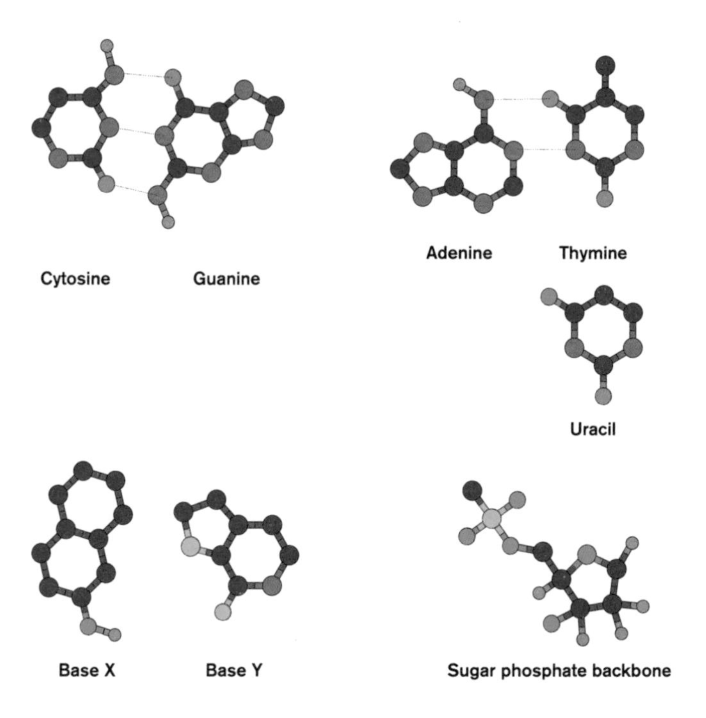

Opposite: CRISPR-Cas9 is a gene editing technique (top) that allows scientists to use custom guide RNA sequences to target particular DNA sequences in the genome and edit or insert genes. The CRISPR system was originally isolated from bacteria, where it forms part of a natural immune response to viruses (bottom). When a virus invades, a small chunk of its DNA may be copied into the CRISPR region of the genome. This sequence then becomes part of a CRISPR-Cas RNA-protein complex that will target and destroy new viral invasions with a matching sequence.

Above: In addition to editing the genetic code, scientists have started to expand the language of life by adding new letters to the DNA alphabet. In 2017, researchers added two synthetic nucleotides: X and Y, to the A, C, G, and T of the normal DNA code to create semi-synthetic organisms—bacteria carrying a six-letter genetic code. As it becomes easier to read and write DNA, perhaps this technology will soon become a significant part of our culture—where not just hard data, but art, graffiti, even espionage is conducted in DNA form.

genome in a predictable place. Modifications to the CRISPR-Cas9 system—replacing the DNA-cutting enzyme with ones that can make specific changes to individual base pairs, changing C to T for example—has allowed precise, single-nucleotide edits to the genome for the first time.

Programming chemistry

Simply by copying genes from one organism to another, we have already achieved some impressive results. The human insulin gene—located on chromosome 7—was cloned and successfully inserted into an *E. coli* cell in 1978, allowing large quantities of the hormone used to treat diabetes to be manufactured in vats of bacteria, rather than extracted from the pancreas of cows and pigs. There are now more than 300 such products—from biopharmaceuticals, including antibodies, hormones, and growth factors, to food and beverage ingredients, biofuels, textiles, and industrial chemicals—being produced by engineered microbes. Genetic engineering is also used to make plants resistant to emerging diseases or improve drought tolerance, to increase the nutritional value of staple foods, or to imbue them with natural insect repellents that have substantially reduced the need for chemical pesticides.

But only so much can be done with isolated genes. Much functionality within living organisms arises from genetic circuits—networks of genes that interact to perform molecular algorithms. French scientists François Jacob, André Lwoff, and Jacques Monod won the Nobel Prize in Physiology in 1965 for discovering a network of genes in *E. coli* that allow the microbe to sense, move toward, and consume lactose—a form of sugar found in milk—when their preferred food source, glucose, is not available.

Genetic circuits are groups of genes that operate together to allow organisms to respond to external stimuli. Like electronic circuits, genetic circuits perform comparisons and execute logical operations within a cell, allowing cells and entire organisms to respond to their environment with preprogrammed logic. These genetic circuits have basic components that act like switches and resistors in electrical circuits. For example, small sections of DNA called promoters and repressors next to genes act to switch gene expression on or off, depending on whether a transcription factor binds. Indeed, promoters and repressors also exist for the expression of transcription factors, which can assemble into chains to control complex cascades of cellular activity and can trigger widespread changes in the chemistry being performed by the cell.

There are thousands of examples of genetic circuits in nature, from the light-sensitive daily metabolic rhythms of plants and animals to the chemical gradients that determine the shape and features of all multicellular organisms and the products and materials we derive from them. As Joe Isaacson, vice-president of engineering at the biotech company Asimov in Cambridge, Mass., observed: "every single thing that civilization sources from biology—food, materials, drugs—was built by nature using genetic circuits to exert fine spatiotemporal control over biochemistry."

Biological engineers have already begun designing their own genetic circuits. For example, in 2000, scientists in the U.S. created a synthetic genetic circuit that acts as an oscillator—periodically producing green florescent protein—making it the genetic equivalent of Prigogine's Brusselator (page 98). Scientists have now replicated many of the functions that allow machines to perform specific algorithms, such as logical expressions ("AND," "OR," "NOT") and arithmetical calculations (addition, subtraction, division).

These building blocks offer new opportunities to create entirely new synthetic organisms, filled with complex, interlinked genetic circuits that perform the precise chemistry we want to achieve, and even interact with other synthetic microbes to perform complex operations. Engineering genetic circuitry promises to make our industrial chemistry, and our crops, more efficient. For example, researchers in the UK have successfully added a synthetic genetic circuit for capturing nitrogen from the atmosphere—based on the genetic circuit used by nitrogen-fixing bacteria in the soil—and inserted it into crops like barley (*Hordeum vulgare*). These plants, if used commercially, could substantially reduce our dependence on artificial fertilizers and reduce the impact of agriculture on the global nitrogen cycle and CO_2 emissions.

Creating new kinds of biological system

A wide range of techniques exist to modify the genetics and understand their impact on organisms. But to move from isolated genetic engineering projects to routine, systematic, industrial synthetic biology, Professor Drew Endy, a bioengineer, argues three new elements of technology are needed: standardization, decoupling, and abstraction.

Standardization underpins the modern world—from screw threads to paper sizes and from internet routers to spaceship docking ports—standards enable collaboration and interoperability. Fortunately, some standards in synthetic biology have already emerged, such as the BioBricks Foundation assembly standards. But a lack of software and hardware

Genetic circuits

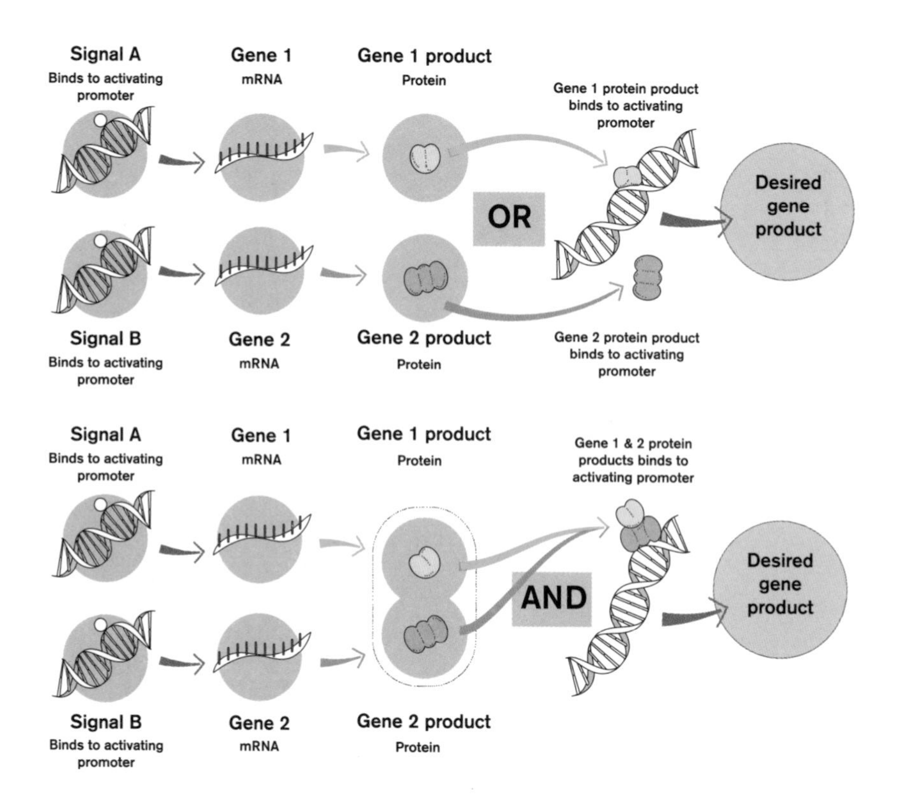

Above: Genetic circuits combine basic components like promoters, repressors, and transcription factors to create feedback loops and logical calculations that allow cells to respond to their internal and external environment. For example, cascading chains of signaling molecules can create "AND" (bottom) and "OR" (top) functions, commonly used in programming, to regulate complex chains of biochemical reactions within a cell.

standardization inhibits the industry from self-assembling into effective ecosystems. Some companies have opted to develop their own bespoke hardware for their software platform to control, while others are challenged with developing more versatile software that can communicate with all the types and models of their customers' hardware.

Decoupling breaks complicated problems into simpler problems, allowing people with different skills to work independently and integrate modules into a functional whole later on. Separating the design of synthetic biology from the fabrication is one such example, and many companies are working on decoupled problems—designing functional genes, circuits, and organisms, and manufacturing the DNA to build them—separately.

Finally, abstraction—familiar from computing and the internet—is necessary to organize different components of synthetic biology into a hierarchy of smaller modules. DNA and the proteins they encode form the lowest layer of a hierarchy that controls whole organisms. We have seen that genes group into genetic circuits that co-evolve to produce efficient biochemical processes like photosynthesis or lactose digestion. In multicellular organisms, these circuits are grouped into systems—the nervous system, endocrine system, respiratory system—the next layer of organization in the biochemical hierarchy. If we zoom out to a whole ecosystem view, these systems group into individual organisms, species, and finally trophic levels or functional groups of an ecosystem.

This system of hierarchical organization, in which smaller subunits group together into higher-order networks, is reminiscent of how the programming of computers is broken into a hierarchical "stack" of modules. A common example in web development is a software bundle known as the LAMP stack, an acronym of its components: Linux, Apache, MySQL, PHP. In this stack, Linux is the operating system, Apache is the web server software, which manages the interface between the computer and the server hardware, MySQL manages the data accessed by the browser, and PHP programs the behavior of the browser app itself. So, our web developer, in theory, need only understand PHP to successfully make an app. Similarly, in synthetic biology biochemistry could be programmed without manually editing each base pair—the equivalent of writing a high-level language rather than attempting to write machine-understandable code directly.

The beauty of the stack design is that it hides complexity within nested modules, making seemingly enormous tasks more manageable and allowing people and organizations to specialize in developing technologies on different layers. Each layer of this stack could just as easily be filled by another technology that performs similar functions, exactly as an organism's functional role in an ecosystem could be replaced by another species that fills the same niche.

Indeed, a famous April Fools' Day prank in 1990 submitted a spoof proposal for a new protocol called "IP over Avian Carriers" (IPoAC), which proposed to replace the physical layer in a computer network stack with a carrier pigeon (*Columba livia domestica*) instead of Ethernet. However, this protocol was implemented in real life in 2001 and is perhaps the first real example of a stack involving a biological component.

If we are to really master synthetic biology to control our manufacturing, waste, and energy systems, we will need to develop stacks of genetic technology that enable the expertise required to create synthetic ecosystems to be broken down into more manageable parts. As an early step on this journey, in 2016, researchers at MIT developed the first programming language for bacteria, based on Verilog—a common language used to program computer chips—with which they created the largest synthetic biological circuit ever built, some 12,000 base pairs long.

Designing life

So far this chapter has shown how genetic engineering has laid the foundations for the first layer, and synthetic biology adds three new techniques that will complete the stack. In the biological reagents layer there is already a rich selection of building blocks to work with, such as annotated libraries of functional genes, proteins, and RNAs; catalogs of functional components like promoters and guide sequences; and model synthetic organisms.

In the second layer, a custom gene is assembled using these basic biological components, perhaps via the DNA printing method we encountered earlier or one of a plethora of other innovative methods. For example, one San Francisco-based biotech company, Twist Bioscience, can produce large volumes of DNA rapidly and cheaply using technology based on a technique we met in chapter 3: photolithography. Silicon plates just 600nm across are etched using photolithography to create 10,000 tiny wells, each of which can hold one custom DNA strand under construction. These custom DNAs are being used to modify bacteria and plants in basic research, drug discovery, biotechnology, and agriculture.

The next layer in our synthetic biology stack is the design step. Just as parts are designed with computer software, synthetic biologists are increasingly using computer-aided design (CAD) and manufacture (CAM) to automate and abstract their design work from the

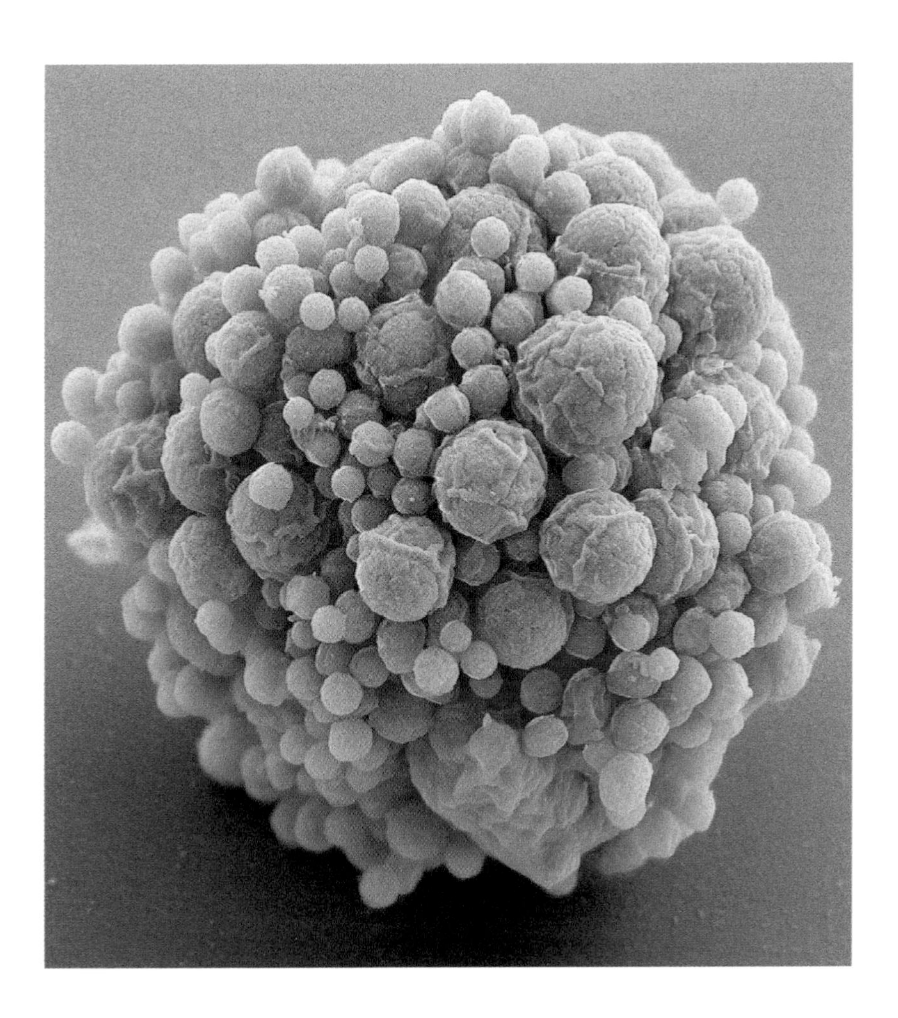

Above: Synthetic biologists have created JCV-syn3.0, a single-celled organism with the smallest functional genome possible. By stripping a real organism down to the most basic components required for life, scientists hope to create the blueprint for a biological chassis that could carry all sorts of natural and synthetic gene circuits.

manufacture of the building blocks themselves. Within this layer, designers can work with functional modules from the reagents layer to assemble new genes, genetic circuits, or CRISPR guide sequences to perform prescribed functions. In 2016, synthetic biology company Asimov unveiled CELLO, a "genetic circuit design automation platform"—CAD software for genetic circuits—that sped up the design process to such an extent that they were able to build 52 new genetic circuits in just one week.

Harnessing genetic circuitry in commercial synthetic biology—the final layer in the stack—presents huge technical challenges. New gene-editing techniques and machine intelligence are assisting scientists in designing and testing their own genetic circuits in living cells for application in a range of different commercial areas. Ginkgo Bioworks, a biotechnology company based in Boston, Mass., describes itself as "The Organism Company," offering custom, genetically engineered organisms for the fragrance and food industries. By inserting genes from plants into yeast cells, Ginkgo has created organisms capable of producing scents and flavorings. Much of the process is automated—performed by robots controlled by computers—which allows the company to achieve the scale and efficiency needed to be commercially viable.

Only with this stack in place can we start to think about engineering entire synthetic ecosystems. Genetic circuits and organisms could be designed to fit into the wider ecosystem of synthetic biology companies and natural nutrient cycles, just as the chemistry of life has been honed over millennia to replenish its own sources of food and fuel. The first steps toward this possibility are already being developed—clock-like circuits that display cyclical behavior have been constructed by engineering signaling between cells within a bacterial population, and oscillating patterns of fluorescent colors have been created using two synthetic strains of *E. coli*, each possessing half a genetic circuit.

Embracing the future

We invite individuals to weigh for themselves the relative merits of the approaches we have explored here. On the one hand, changing biological genes may create more food and medicine for humans, bacteria for digesting waste plastic, plants and microbes to help recycle landfill into useful materials, and create routes to manufacturing new materials that reduce emissions and help mitigate climate change. But it could also have unintended evolutionary consequences as well as long-term unexpected impacts on the biosphere.

The SynBio stack

SynBio stack	Products
Application	Leather, meat, industrial chemicals, fragrances, flavors, drugs, habitats, more...
Bio CAD/CAM	DNA part design software, data analytics tools, LIMS, design of experiment (DOE), job scheduling, more...
Process execution	Liquid handling robots, acoustic dispensing, cloud labs/CROs, microfuidics, electrowetting, more...
Biological reagents	Synthetic DNA, standard parts libraries, restriction & ligation enzymes, model organisms, more...

Above: Companies in the emerging synthetic biology ecosystem can be stacked into functional layers. Layer one, the base layer, would contain the basic biochemistry, libraries of DNA or protein. Layer two would execute the processes —fabricating the DNA. Layer three is concerned with computer-aided design (CAD) and computer-aided-manufacture (CAM) technology, which direct the technologies in layer two to assemble the components in layer one. Finally, above that, would be the technologies that apply the synthetic biology to perform particular functions. Based on an original SynBio stack concept from SynBioBeta.

Individuals and organizations are right to have concerns about our ability to manage the complexity of genetic circuits, and the value, safety, and ethics of such products need to be evaluated rigorously on a case-by-case basis. We must be able to understand genetic systems well enough to use them in a safe, ethical way that doesn't adversely affect humans or other life.

Through responsible synthetic biological research, we may be able to handle complex systems, like cells, populations, and ecosystems, and in the process gain new insights into how to simplify the exchange of industrial waste streams. Such a systems-level understanding of material flows from the micro- to the macro-scale could help us to start living within the immutable physical limits of planet Earth. We also see an opportunity for people to reconnect with nature and relate to it on a new, deeper level.

Combining additive manufacturing with synthetic biology

As well as transforming energy generation, agriculture, pharmaceuticals, and industrial chemistry, synthetic organisms could be engineered to produce feedstocks for additive manufacturing, allowing us to bring farms and factories closer together and to expedite the development of circular economy habitats.

Synthetically grown analogs of a diverse range of materials and products, from meat to industrial chemicals, are already available on the market. ModernMeadow is a manufacturer of biofabricated materials; it designs and engineers living cells that grow collagen—the protein naturally found in animal skin—for assembly into sheets that can be tanned and used like leather. Another example is Bolt-Threads, a company that produces artificial spider silks from genetically engineered yeast. Materials like these might enable us to eliminate huge tracts of farmland currently used to grow textiles like cotton. Given the tensile strength of spider silk (page 96), such synthetic materials could even replace energy-intensive building materials like steel.

Scaling up these ideas, the ability to reconfigure our factories to make any object or material, without having to redesign the infrastructure every time, would be remarkable. Imagine being able to manufacture a functioning product like a smartphone, or food, or clothing, from a single supply of nutrients entering a building through a pipe. Printeria, a bioengineering device no bigger than a domestic microwave, may be the first step on this journey. It can print genetic circuits in bacteria by using nano-sized droplets that can be moved about to perform the steps needed to assemble custom plasmids and insert them into bacteria. Technological engineering like this paves the way for a future where entire tissues and organs could be printed, with exciting prospects for transplant medicine, and indeed the food industry.

Such a universal and reprogrammable production process may enable a remarkable step change in industrial operations, allowing companies to exchange products and waste materials more efficiently. If companies could use synthetic biology to carefully design waste streams—as well as their products—so that their outputs were beneficial to other companies, it could lead to the emergence of natural reservoirs of stock materials that could be shared universally. The circular future is already starting to take shape.

Making materials smarter

At this point, we have seen how biologically inspired chemicals designed using synthetic biology and additively compiled into bulk materials could help us move closer to a circular economy. These materials would be able to replace any material that we currently rely on, from steel to cotton. But these new manufacturing techniques would also open new avenues for us to develop more advanced materials and systems that could perform entirely new functions and roles, which we might one day be able to program digitally via home computers.

In chapter 6 we explore how synthetic biology and additive manufacturing lend themselves to the development of a whole catalog of different smart materials.

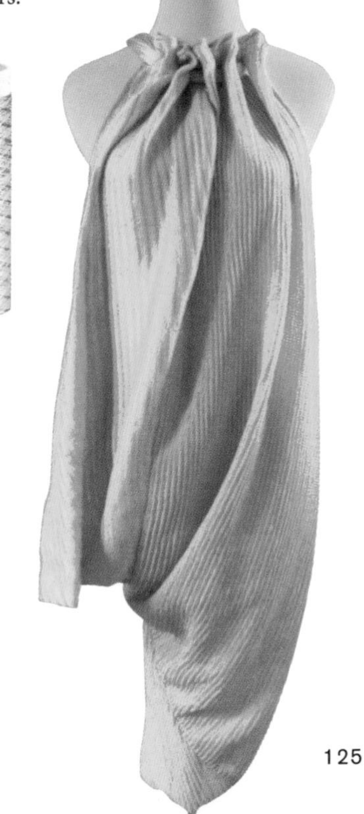

Right: US company Bolt Threads' trademark Microsilk™ is produced using genes copied from spiders and inserted into yeast. It has been used to create commercial clothing as well as works of art, such as this Microsilk dress designed by Stella McCartney, which was displayed at the New York Museum of Modern Art (MoMA) in October 2017.

Storage and memory

DNA offers the potential for a far higher data-storage density than any electronic hard drive humans have invented to date (*Living hard drives*, right). Thus, our biosmartphone could store vastly more data than a modern smartphone in a much smaller space, but it might also enable us to store the smartphone's manufacturing instructions within the phone itself, just like a living organism holds all the instructions needed for reproduction and development.

CRISPR is a form of genetic memory, making it the perfect candidate for storing our own data in strands of DNA. Although only recently harnessed by humans, the CRISPR system was developed billions of years ago, and acts like a memory of infections past, capturing viral DNA and storing it in the genome, archiving it ready for the next invasion. In 2017, Seth Shipman used CRISPR-Cas9 to store a file containing an animation of a famous series of photographs by Eadweard Muybridge depicting a galloping horse. Each pixel of the animated image was translated into a letter in the DNA code, and the sequences were handed to the bacteria frame-by-frame, ensuring CRISPR stored the animation in the correct order. Such an achievement shows a glimpse of how we might read and write data to our biosmartphone's hard drive, but many challenges remain.

One hitch is that DNA is a one dimensional "tape" of information—making it a comparatively slow process to access genes for reading or writing. Current research is focused on reducing the error rate, improving genetic sequencing and synthesis techniques, and developing more precise methods for targeting specific sequences. Recently, researchers at the University of Cambridge in the UK developed a DNA hard drive made of single-stranded DNA, with a unique, targetable address for each gene and a system for data encryption.

The great advantage of such a bacterial hard drive over traditional archival storage is that nature has evolved over millennia to be extremely good at maintaining the data stored in DNA faithfully. Even many generations later, the data recorded in bacterial genomes has not degraded. Furthermore, DNA is a very stable, inert molecule when kept in a cool, dry environment. And, unlike the silicon chips that store data in a real hard drive, DNA can be manufactured at ambient temperatures with simple organic molecules. Indeed, the use of DNA as a medium for short-term

Living hard drives

Scientists developed techniques to synthesize short sequences of DNA in the late 1970s and a decade later they were already using it to store their own files. In 1988 the first non-biological data—an image of an ancient Germanic rune—was encoded in DNA and stored in *E. coli* by artist Joe Davis and Harvard Medical School researcher Dana Boyd. Scientists have gone on to store Shakespeare's sonnets, an audio clip of Martin Luther King Jr.'s "I Have a Dream" speech, a computer operating system, and a pdf file of James Watson and Francis Crick's paper on the structure of DNA, all within the reams of DNA. Sequencing allows us to retrieve the data—creating the first biological hard drives.

In chapter 4 we saw that a "hard drive" made of biological cells could potentially hold around 1.4 million terabytes of data in DNA—more than enough to store a movie of your entire life. To date, the largest amount of data stored in DNA was 200 megabytes, but just as the number of transistors on a computer chip increased from a few thousand to 50 billion in a matter of five decades, DNA hard drive technology too is likely to continue to grow exponentially. Transitioning to large-scale DNA-based data storage may be inevitable, as the accelerating growth of humanity's data runs into the physical limitations of silicon chips.

and archival storage is a subject of active commercial research. With a DNA hard drive for long-term storage, the short-term working memory of our biosmartphone could be RNA, in which short sections of recently copied mRNA could float around in the biosmartphone milieu acting as a temporary memory storage for shorter time-scale molecular processes to work on.

The idea of using human designed molecules, rather than DNA, to store information is also being researched, but so far protein and DNA are the molecules of choice because the machinery already exists in living organisms to exploit them. Thus, synthetic biology may be the perfect tool in which to develop storage systems that take advantage of the colossal information density achievable in molecular architectures.

6 Emergence

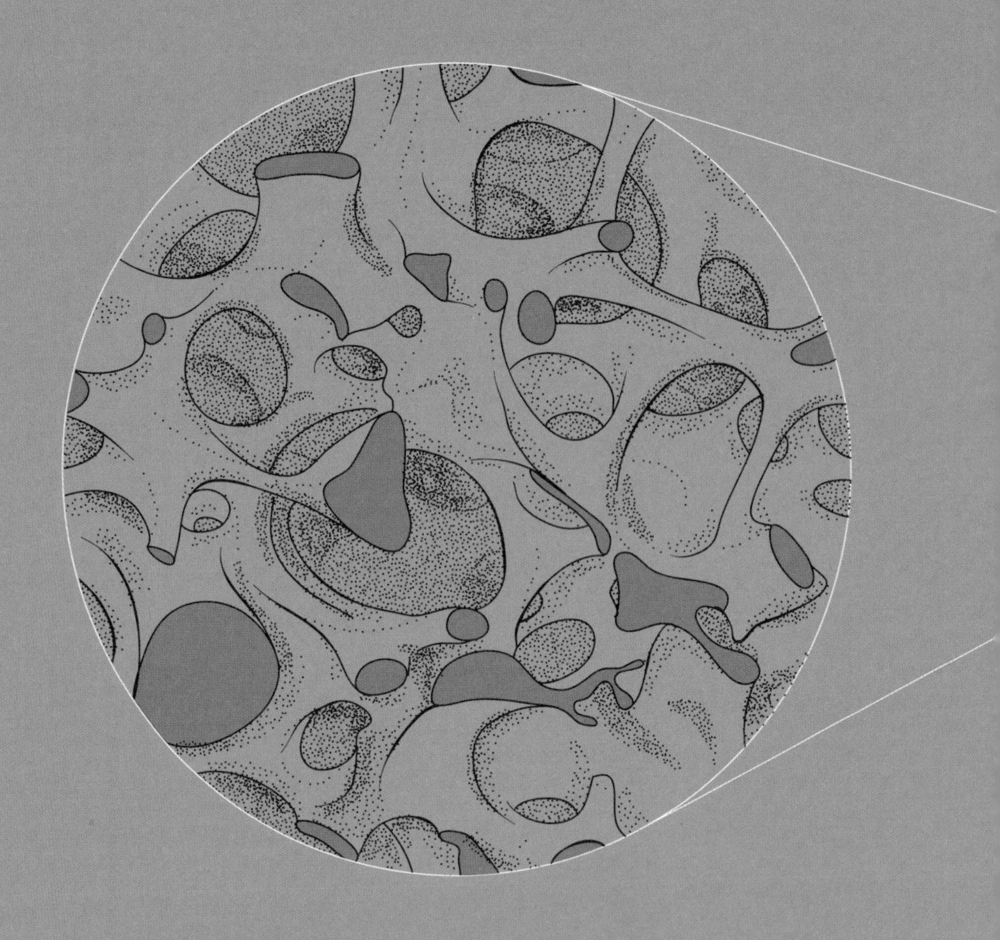

"In the case of all things which have several parts and in which the totality is not, as it were, a mere heap, but the whole is something besides the parts, there is a cause."

Aristotle, *Metaphysics*, Book VIII, translated by W.D. Ross (1908)

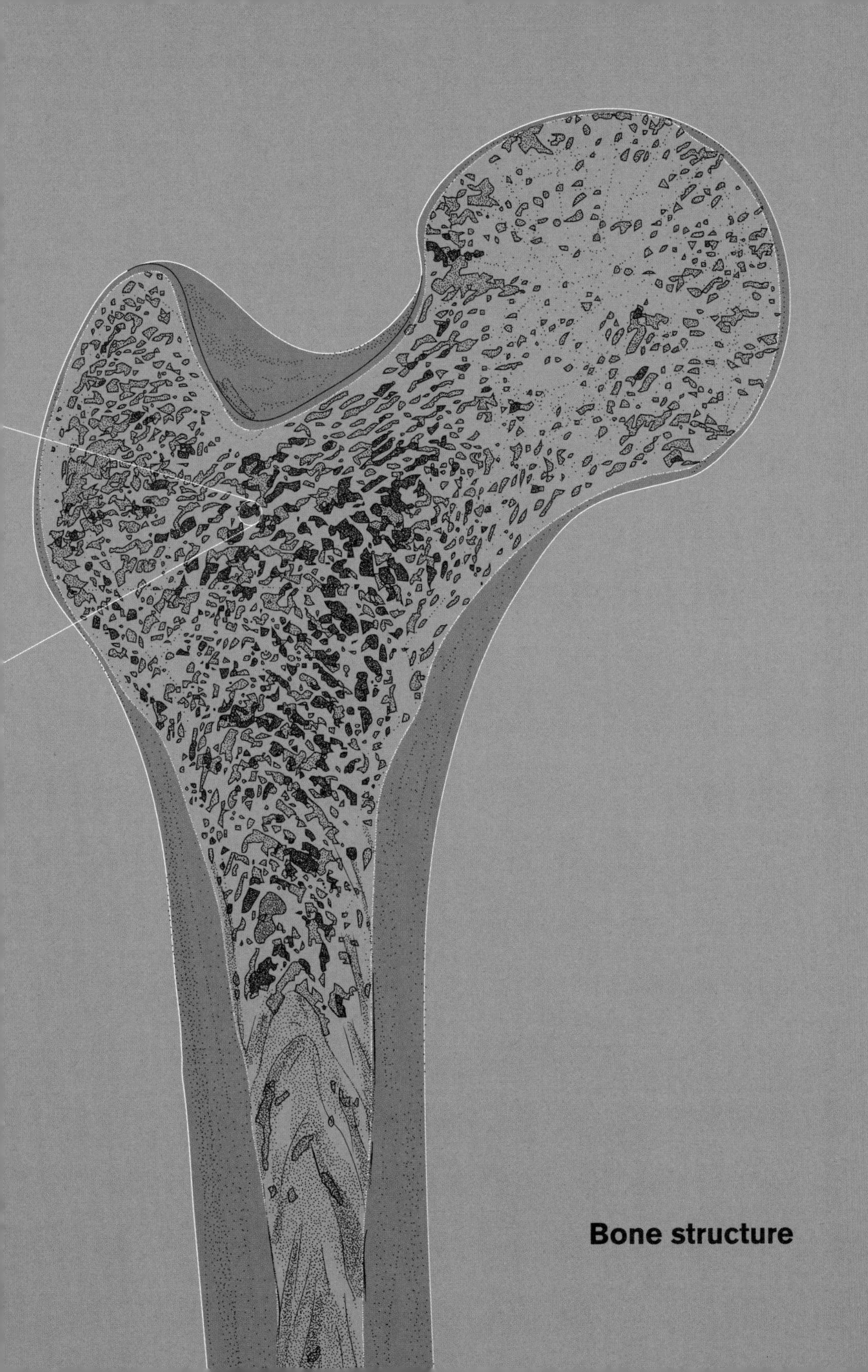
Bone structure

On our journey to circularity we have encountered two extremely powerful technologies. Synthetic biology has the potential to create a huge range of molecules on demand, and additive manufacturing—whether 3D, 4D or dissipative—can assemble many of these molecules into arbitrary macroscopic shapes. Put them together and a profound and mighty possibility emerges—complete local computer control over every aspect of our manufacturing—all the way from molecules up to ecosystems.

This vision is beguiling when presented in such an over-simplified way; the devil is in the detail. Many colossal technical challenges are embedded within this simple idea, not least of which is the vast gulf of scale and complexity that exists between molecules and the macroscale products we use every day.

To understand the pitfalls and possibilities as we make this leap across scales, we first need to learn about the idea of emergence. Complex systems—such as the human body, ecosystems, biological materials, the internet, honeybee colonies, and even fundamental physics—all display so-called emergent phenomena: behavior that cannot be seen in the individual components from which the system is made. Having understood emergence, our goal then becomes to combine synthetic biology and additive manufacturing so that the circular economy is an emergent property of the synthesis of these two technologies. We have powerful tools at our disposal, such as the technology stacks discussed in chapter 5, but we must learn to use them masterfully to construct a circular economy.

Leaping the divide from molecules to materials

By finding combinations of components, materials, and systems that work together we may be able to fashion tools like synthetic biology and AM into foundational technologies for both halves of the circular economy—technical and biological. Together these tools could add considerably more value than they provide alone—emergent value created by the fusion of two powerful technologies.

The smaller the palette of materials our combined technologies use, the easier it is to build a circular economy. If we could systematically find and control the emergence encountered by our systems, we might not have to introduce new materials for every new function we want. Instead we can be clever about combining base materials to achieve new effects.

In the biological half, most of the job has been done for us. Just four base pairs and 20 amino acids (page 93) are the basis of a planetary wide network of ecosystems that can use sunlight as a power source to perform incredibly complex additive manufacturing at the molecular and indeed cellular level.

Perhaps if both halves of the circular economy were based on the same principles then the two flows of matter could be made completely compatible, and we could transplant materials, processes, or functionality between them more easily. For instance, could we use farms of engineered crops to extract metals from landfill instead of mining raw materials from ore? Could we combine synthetic biological cells to form fully functioning devices as well as manufacturing chemical feedstocks? Even if we don't end up using the same base materials in both halves, understanding how emergence arises in biological systems will give us a few ideas as to how to develop the technical half of our economy. Whichever strategy we end up using, we have a great deal to learn from the biological loop about using emergence to bridge the gap between the molecular and macroscopic scales.

The evolution of multicellularity

In the natural world, the gap between molecules and macroscopic functioning systems is bridged by the cellular architecture of organisms. Most of that complexity is managed within individual cells, which arrange hundreds of trillions of atoms into a reaction network consisting of around 10,000 broad classes of molecule. The details of these reaction networks differ between types of cell. However, there are only around 200 different types of cell in a large multicellular organism, and those cells group into a mere handful of different organs and tissues. Therefore, although they are highly complex, organs, tissues, and multicellular organisms are significantly less complex than the operations *within* a single cell.

Synthetic biology gives us the tools we need to manage individual cells—what remains is to understand how cells work together. And when cells cooperate, remarkable larger-scale phenomena emerge, such as moving muscles, networks of electrical impulses, and hierarchical materials like bone or wood.

The hallmark of materials built by biological cells is that the structures possess sub-structure, and the sub-structures possess another level of sub-structure, and so on until we reach the molecular scale. Clearly such clever materials fabrication techniques are of relevance for AM and all kinds of applications in the technical loop, but before we learn how to build such materials and tissues with engineered cells, we must first understand how a collection of cells exhibit emergent properties not displayed by individual cells and in doing so we unravel the very reason that multicellular organisms evolved in the first place.

Introducing *Volvox carteri*, one of the simplest multicellular creatures on Earth, which live in colonies that form hollow spheres. The cells on the outer surface each have a whip-like appendage used for swimming known as a flagellum, making the colony as a whole capable of synchronized motion that propels it through the water.

Volvox carteri is one member of a group of microbes—the Volvocales—that include free-living single-celled species, colonial species that group together into balls of a few hundred identical cells, and simple multicellular species that form colonies of 1,000 to 50,000 cells, which are differentiated into two or more different cell types. The larger colonial species struck upon an emergent benefit of cooperative behavior: division of labor and specialization.

V. carteri cells on the outside of the colony grow flagella like their single-celled relatives, but they can no longer reproduce; the reproductive function is reserved for cells on the inside of the colony, which can no longer swim. By dividing labor, cells can specialize and become highly efficient at their role. For example, single-celled species have to shed their flagella to reproduce, but by separating the swimming and reproductive roles into different cell types, large colonies can perform all these functions *simultaneously*.

Colonies are also able to feed and reproduce more efficiently than their single-celled relatives. The specialized swimming cells in *V. carteri* colonies don't just swim through the water to guide the colony to food, they actively work together to guide the local flow

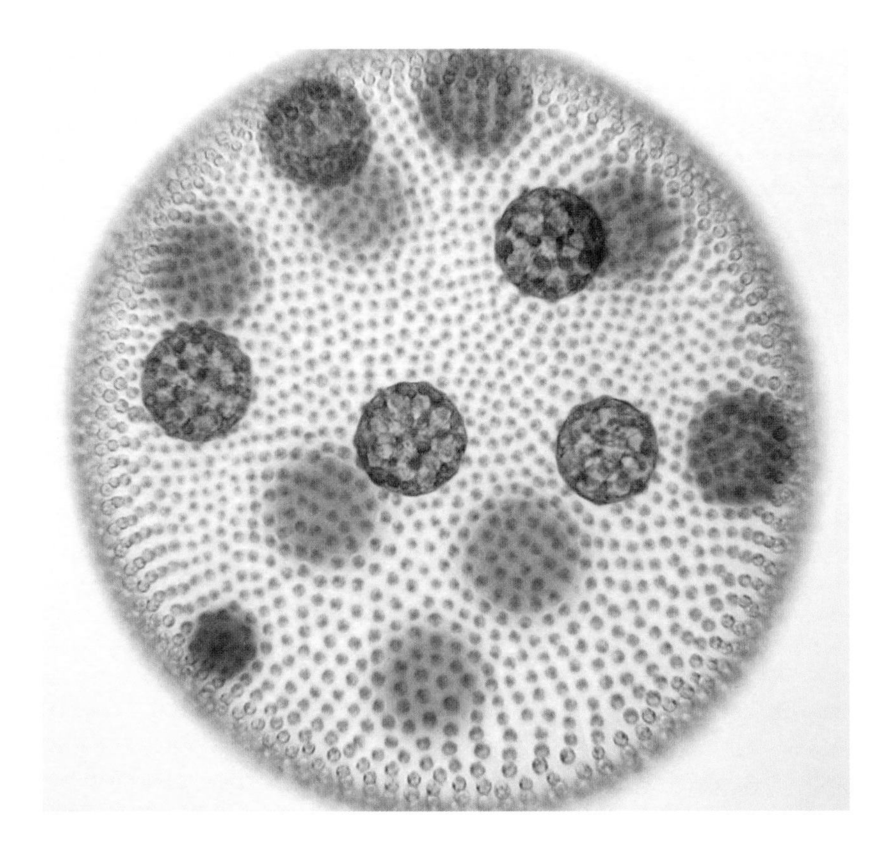

Above: At the scale of individual cells and small colonies, the water is like treacle, making it very hard to swim, so hungry microbes have to wait for food to flow past by diffusion. *Volvox carteri* colonies (pictured) combine the swimming power of many cells to travel further through the treacle to where the food and light is, and fluid dynamics cause water to flow more easily over larger scales, so each cell within a colony has more access to food.

The flag model of embryonic development

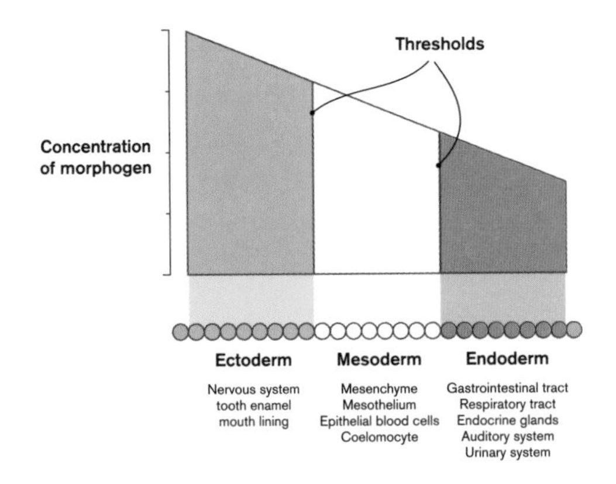

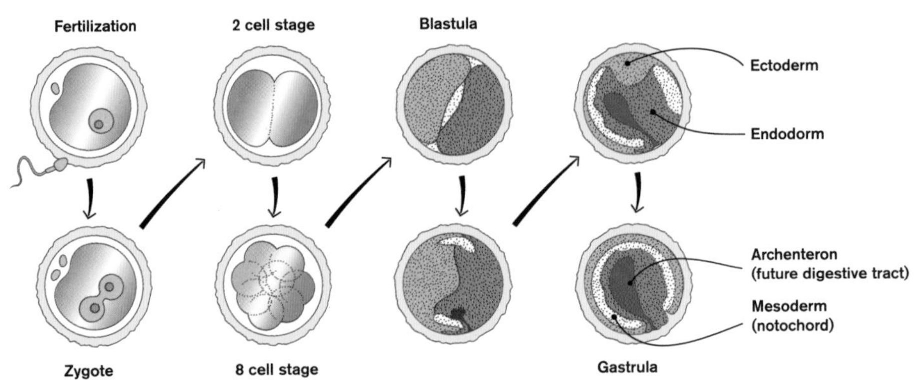

Above: The different cell types in the body are achieved using gradients of chemicals in the egg cell called morphogens, which are often transcription factors. As the egg cell repeatedly divides during embryonic development, the initial gradient means cells contain different amounts of each morphogen, and the concentration determines the fate of the cell. In this example, cells with a high concentration of the morphogen become ectodermal cells (blue), while a low concentration produces endodermal cells (red). Cells with an intermediate concentration become mesodermal cells (white). Different cell types express different genes so they have a specialized morphology and can perform distinct chemistry—two emergent properties of multicellular organisms.

of water currents to bring the nutrients to them. There are also many more swimming cells than reproductive cells, allowing the colony to acquire enough nutrients to grow very large, and exploit changes in the physical properties of water that occur at larger scales to swim more efficiently. The swimming cells take up nutrients from the environment by diffusion and then pass most of them onto the reproductive cells, which are expending a lot of energy and resources on reproducing. This resource sink within the colony maintains a steep chemical gradient between the swimming cells and the external environment that draws nutrients into the colony more rapidly.

From a single-celled perspective, the emergent behavior of colonies of *V. carteri* is possible because the two cell types activate different genes using transcription factors (page 118). Both daughter cells have an identical collection of genes—their genotype—but they activate different sets of the available genes creating distinct appearance and behavior. In fact, all multicellular organisms use transcription factors to create hundreds of different cell types from a single fertilized egg. Known as the flag model of embryonic development, gradients of transcription factors in the fertilized egg are passed down to the daughter cells, triggering suites of genes that give each cell its specific behaviors. Therefore, the appearance and behavior—the phenotype—of different cell types within a multicellular organism are an emergent property generated spontaneously from chemical gradients in the developing embryo.

Natural selection has favored genes for colonial living because they give the colonies of algae better hydrodynamic control over their environment and location, as well as better nutrient uptake, compared to single-celled algae. These remarkable algae demonstrate how groups of cells working together can take advantage of dissipative phenomena (page 98) to create emergent behavior that cannot be produced by single cells. The result is multicellular life.

Over the course of evolution, it's not just groups of cells that have discovered emergent behavior. Groups of organisms have discovered the emergent benefits of collaboration, from the intricate patterns of flocking birds and shoaling fish that dazzle predators, to the collective intelligence displayed by social insects. Colonies of ants, bees, and wasps share many characteristics with individual multicellular organisms: they reproduce as a single unit; individuals within the colony specialize in certain tasks; and they all share a near-identical genotype. For this reason, some scientists have dubbed social insect colonies "superorganisms."

Like multicellular organisms, superorganisms create different phenotypes from a single genotype using chemical gradients that trigger genetic circuits. And, like hierarchical materials, superorganisms display impressive emergent properties as a result of the complex interactions between different levels of organization—the cell, the organism, the colony.

Hierarchical materials

This monumental transition from single to multicellular life has occurred many times among different kinds of organisms. The first occurrence around 3 to 3.5 billion years ago was for cyanobacteria, for Eukaryotes it first occurred around 1 billion years ago, but the biggest explosion of multicellular life occurred between 600 to 700 million years ago, and since then they have gone on to find incredible ways of producing bulk materials by combining teams of specialized cells in joint additive and subtractive manufacturing processes.

For example, three types of cell work together to produce bone: osteoblasts, osteoclasts, and osteocytes. Osteoblasts fabricate new bone, osteoclasts reabsorb it, and osteocytes, embedded throughout the bone matrix, chemically control the other two cell types in response to local stress and strain signals in the bone, and global signals in the blood.

Bone is a composite material formed from a softer cartilage component and a harder mineral component. The cartilage is made from a protein called collagen (the most common protein your body produces), and the harder component is made from a crystalline form of calcium phosphate mineral. The hierarchical composite structure gives bone a fracture resistance that exceeds that of either collagen or calcium phosphate. The toughness of bone is therefore an emergent property.

At the molecular scale, three collagen molecules twist together to form a short, helical bundle called tropocollagen. This rod-like structure is the building block that can stack up to form much larger fibers, which twist up into even larger bundles to form the cartilage that defines the overall shape of the bone.

The calcium phosphate is deposited onto the cartilage in small crystals called platelets, alongside other small molecules, such as citrate and water, which separate layers of mineral and ensure that each platelet layer grows parallel to the collagen. Larger, single crystal platelets would be prone to breaking. Instead, the citrate-filled gaps improve the mechanical properties of the material by creating a "region of disorder," that likely helps stop small cracks from spreading. The result is a hierarchical material that is strong, lightweight, flexible, thermodynamically viable, and inherently recyclable, and for which—over millennia—we have found many uses, from needles to teaspoons.

Plants, too, have utilized emergence to produce materials with some helpful properties. For example, pinecones close their scales when exposed to moisture, and reopen when they dry out. This environmentally-triggered emergent phenomenon is achieved by combining two materials in a single plate—one that is hygroscopic, meaning

Above: The apparent intelligence of an ant colony emerges as a result of interactions between many individuals, each following simple rules such as "lay a pheromone trail if you are returning to the nest with food" or "if you are not carrying food, follow a pheromone trail." Evolution has honed these rule sets to produce emergent behavior that is effective for finding food, defending the colony, and reproducing. Leafcutter ants such as *Atta cephalotes* (pictured here) use their powerful mandibles to harvest small pieces of leaf, which they use as fertilizer for an underground fungus farm. Their emergent foraging trails can remove 220lbs (100kg) of vegetation per hectare each year.

it absorbs water and expands, and one that isn't—causing the plate to bend when wet, closing the cone and protecting the delicate seeds inside. Such materials are known as smart materials because they are able to respond dynamically to their surroundings.

A more dramatic example comes from *Bauhinia variegata*, a flowering plant that deploys a process called explosive dehiscence to fling its seeds almost 200ft (60m) from the parent. The ingenious method takes advantage of the emergent properties of interactions between diagonally placed cellulose fibers that form the seedpod. The long, thin specialized cells are reinforced by wooden material along the cell walls that makes them rigid.

When what is made is more than the sum of its parts

A particularly beautiful example of emergence is the synchronization of the oscillating lights of glowworms. Without leaders or commands, the whole population synchronizes its famous bioluminescence. This synchronicity is thought to amplify the glowworms' collective signal, helping every individual attract potential mates.

But what *is* emergence? While the advantages of emergence are clear, it is very difficult to define; a full exploration requires its own book to do it justice. In the same way that physicists have battled for more than a century against the seemingly impenetrable fundamental principles of non-equilibrium physics and definitions of information that we briefly discussed in chapter 4, philosophers have long debated the real definition of emergence to no avail. In fact, as we will see later, these conundrums may all be related.

Definitions of emergence have existed since the time of Aristotle, to whom the famous phrase: *"the whole is greater than the sum of the parts"* is often attributed, but as the translation at the beginning of this chapter indicates, this could be a misquote. Ross's translation implies Aristotle thought that when combinations of components occur something *new* might be created that is there *"besides the parts."*

Such a concept is known as strong emergence—the idea that as components are connected into systems new properties emerge that in some sense have their own existence. Some candidate examples include the detailed properties of large bodies of water that have eluded simulation based on the component water molecules; consciousness itself may be an example of strong emergence; and of course quantum entanglement—in which two particles are deeply connected making their internal states mutually dependent. In contrast, weak emergence refers to the idea that emergent phenomena can always be reduced to a set of underlying rules, and fully explained in terms of the constituent

Above: *Caesalpinia pulcherrima* (left top and bottom) and *Calliandra surinamensis* (right top and bottom) both use explosive dehiscence to disperse their seeds. As the seedpods mature, their fibrous shells dry out and shrink in the direction of the fibers. The fibers in each layer are at right angles, so each half of the seedpod tries to warp in the opposite direction, but the two halves are held firmly shut. Elastic energy accumulates within the fibrous husks and when it exceeds a certain threshold, the pods spring open—with a characteristic clacking sound—releasing all that pent-up energy to fling seeds in all directions. With this basic design, different orientations of the fibers—diagonal or lengthways down the seedpod—cause the seedpod to spring into different final shapes, such as spirals (left) or loops (right).

parts. In weak emergence there is no "besides." We could simulate the components and rules of a system and the same emergent behavior evident in the real world would emerge in the simulation. For example, a flock of starlings creates a murmuration, which takes on a mesmeric life of its own, but we can simulate this phenomenon by programming rules for each bird, without programming the behavior of the whole flock, step by step, into a computer.

Emergence, mathematics, and the birth of computing

A common trick to grasp a difficult concept is to flip it on its head and understand the opposite idea. Some philosophers view emergentism—a belief in strong emergence—as the opposite of reductionism, which is the belief that a system can always be reduced to its component parts. For example, in the late 19th and early 20th century there was a school of thought that all of mathematics could be reduced to pure logic. However, the existence of paradoxical statements such as "this sentence is false", threw mathematics into disarray and put an end to the idea of reductionism.

In attempting to eliminate such paradoxes from mathematics, work by British philosopher Bertrand Russell and his mentor Alfred Whitehead led directly to analytic philosopher Kurt Gödel's well-known incompleteness theorems in 1931, which proved that it is not possible to prove or disprove every statement that can be written in a logical system by starting from the assumptions, known as axioms, that define the system. On some very intuitive level, Gödel's theorem seems to resonate with the idea of emergence. By saying that some true statements can't be proved from the axioms, it is as if some statements could be written down that exist *beside* the axioms, to blend Aristotle and Gödel's ideas.

Gödel's theorems led directly to the birth of computing, which explains their relevance here. Gödel devised a scheme in which he could replace logical statements with numbers. For example, in the computer language ASCII, the simple logical statement "If a=b then b=a," would be represented as "097061098061062098061097." These encoded mathematical representations of logical formulae are known as Gödel numbers and they were a precursor to computer programs. They inspired Alan Turing, the famous cryptographer and mathematician, to imagine an intriguing hypothetical machine.

Turing's machine has an infinite strip of paper on which it can write down or erase symbols. By following a sequence of simple instructions, like "If the current symbol is a 1 then delete the 1, write 0 instead and step to the right," the machine can follow an algorithm to write down any computable number, which is the set of real numbers that can be computed to any precision by a finite algorithm.

Led by Gödel's numerical encoding scheme, Turing assigned numbers to each logical instruction that could be performed by one of his machines. A string of such numbers—a Gödel number—thus represents a sequence of operations for the machine to perform, which, given a starting tape configuration, will result in a particular number being written on the strip. A computation!

Turing realized that a "universal Turing machine" could be constructed that would take a Gödel number as input and compute the output of the algorithm encoded by that Gödel number. Moreover, the universal machine could take *any* such number as an input and compute the output. Such a device is a computer and the Gödel number encoding the algorithm is a computer program.

It is quite remarkable that our quest to understand what emergence is leads us straight to the idea of computing. Perhaps the same kind of revelation for materials is just what we need to build our emergent circular economy—a system for finding emergent properties from a standard code of chemicals: a chemical computer. We pick up this idea in chapter 8, but first we will consider some of the ways humans can, and might, make use of emergence in the circular economy.

Emergence and non-equilibrium physics

Engineers routinely make use of emergent phenomena. Consider the emergent forces of lift and drag. You could place a cube of aluminum in a wind tunnel and, depending on its orientation, it would experience lift—an upward force from the air—and a lot of drag—air resistance. You could re-form the same mass of aluminum into the shape of a wing—known as an airfoil—and place it into *static* air and the wing would experience no lift or drag at all. To fly—itself an emergent phenomenon—the correct wing shape must be placed in a moving air flow, so as to maximize lift and minimize drag. The pressure differentials on the upper and lower surfaces and the change in momentum of the air flow as it leaves the wing, add up to create an upthrust greater than the weight of the wing, with minimal drag, allowing the aircraft to fly.

The emergent phenomenon of flight only arises when enough energy is being expended—when the system is out of equilibrium. A continuous energy input is needed to maintain these pressure and velocity gradients and keep the system flying. Without a driving force from the engines to overcome drag and generate lift, the aircraft would fall out of the sky, even with the perfect wing. So, we see, at least in this example, that non-equilibrium physics links directly to the emergence of flight. If they are more generally connected that might explain why both concepts have proved so challenging to define.

Attempts have also been made to exploit the emergent properties that can arise in hierarchical materials. For example, metamaterials have properties not found in naturally occurring substances. Typically, such materials interact with their environment under non-equilibrium conditions—they might interact with sound or light energy, or mechanical perturbations such as vibrations or tensile stress. With clever engineering, it is possible to manage the flow of energy through these materials to yield properties that are useful, such as lenses that focus or bend light, sophisticated radio antennas, or shock absorbers. For example, whereas most materials get thinner when stretched, like an elastic band, auxetic metamaterials get fatter, and 3D-printed auxetic materials could be used to create energy-absorbing foams, dilators to open blood vessels, adaptive footwear and clothing, or even self-cleaning filters.

Emulating biology in the technical loop

Although biology principally trades in soft matter, it can also produce hard materials, such as shells, casings and bone, and like all biological materials they are intrinsically recyclable and repairable; properties that have long been sought in engineering circles.

Bone is completely remodeled throughout its existence, and consequently, is also repairable. During bone formation, osteoclasts—one of the three cells we met earlier—dissolve mineral components of the bone using hydrochloric acid, while protein-digesting enzymes break down collagen. During that process, the osteoclasts create little trenches in the bone under instruction from the osteocytes, and they signal to recruit osteoblast cells that add new material behind them. Thus, bone formation is simultaneously additive and subtractive manufacturing—perhaps such a process should be called incremental manufacturing?

Osteoblasts build up the bone behind the osteoclasts and some get trapped in the structure to become osteocytes, the third and most common cell in the team. Once trapped, osteocytes form a network that monitors the local environment throughout the skeleton and controls the teams of osteoclasts and osteoblasts remodeling the bone surface. The cells are able to determine an optimal design using an embedded logic that is distributed throughout the osteocyte network. The main two types of bone are cortical bone and trabecular bone.

Lift in an aircraft

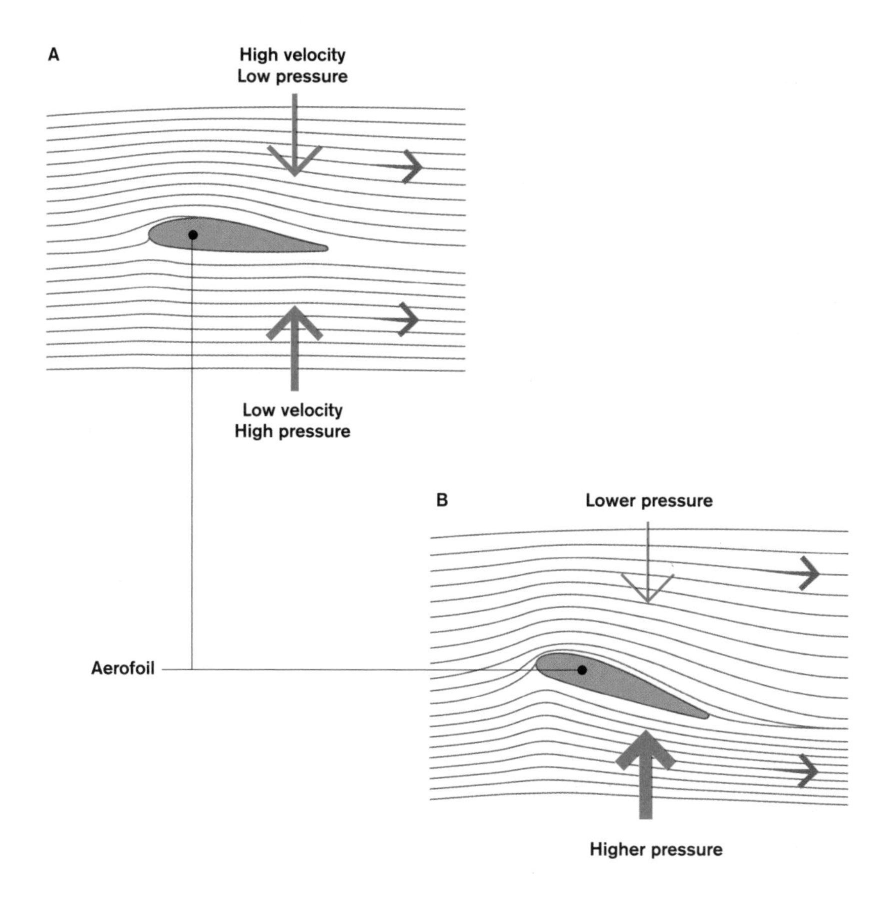

Above: To fly, a wing needs to have a suitable shape and be in a moving stream of air. Engineers use streams of smoke to visualizse the flow of air around a wing in a windtunnel (pictured above in blue). A combination of the shape of the wing and the angle of attack causes the air flow —and therefore the streamlines—to change speed and direction around the wing. The pressure of the air—indicated by the distance between the streamlines—increases below the wing and decreases above it, creating a pressure difference that yields emergent forces such as lift and drag. The forces can be directed up or down allowing aircraft to fly upside-down.

Above: 3D-printed hierarchical materials known as auxetic materials are unusual: stretching the material in one direction causes it to expand in the other direction. Auxetic materials have been used in a variety of innovative designs. For example, this stretchable fabric created by Maria Alejandra Mora-Sanchez could be used to create longer-lasting clothing that adapts to the wearer's body, with obvious applications for infant and maternity wear.

Dense cortical bone forms the outer surface of most of the skeleton (and accounts for most of the weight), while the center is filled with a lightweight but strong latticework of material known as trabecular bone. Similar engineered structures, optimized for strength and minimal resource use, have been designed using generative design techniques (page 87).

Given the sophistication of bone formation, it may not be far-fetched to suggest a synthetic biology stack (chapter 5) in which different cell types work in sequence to manufacture and assemble materials for various products or infrastructure. The groups of cells could be released into a manufacturing chamber sequentially—perhaps a bioreactor similar to breweries, where alcohol is made from sugar—so they wouldn't all have to work together at the same time as they do in an organism. They could then be washed away from the finished product in stages ready for reuse or recycling, or trapped in a final structure to stay there for the operational phase of its life cycle.

There are billions of osteocytes throughout the bone system that signal construction and demolition teams, and, like the synchronized glowworms, the bone cells spontaneously generate an ordered structure that adapts to the environmental stresses and strains that the host organism experiences.

However, to exploit biological processes directly in our manufacturing, we don't necessarily need to build completely new multicellular organisms. Cells have evolved with multiple behaviors that can emerge from a single genotype, and as we saw with *Volvox* earlier in this chapter, multicellular organisms can create a wide array of different cell types even from a single starting cell. Similarly, superorganisms such as ants, with the same genotype, display different suits of behaviors that group them into castes—forager, soldier, nurse—a phenomenon known as phenotypic plasticity. By transcribing different genes to create different sensitivities to certain stimuli we could use the same cell but engineer it to activate different manufacturing tasks at different stages, triggered by a stimulus such as light or chemical changes.

Finding solutions in large search spaces

American theoretical physicist David Bohm suggests in his 2002 book, *Wholeness and the Implicate Order,* that reductionism is less about breaking a system into parts as it is about considering a system from a single perspective. This observation led him to redefine holisticism, which conventionally means to look at all the components of a system together, but Bohm's interpretation is that holisticism is more accurately

Cloaking devices

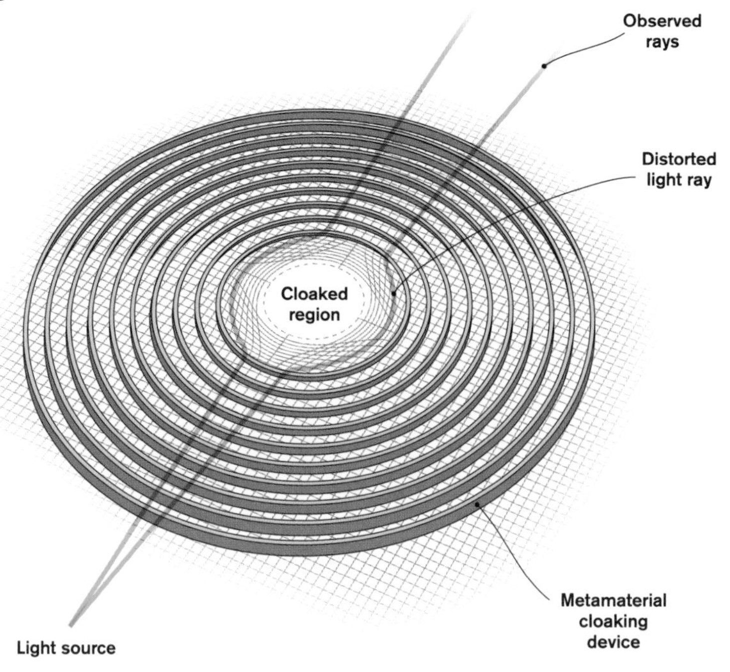

Above: A truly sci-fi example of a metamaterial application would be a cloaking device—a material with a negative refractive index would bend light around a region, thus hiding any objects enclosed within, at least over a range of light frequencies. The light rays entering the cloaked area are diverted around any objects in the central region of the device and returned to their original course. Anyone outside the device would see only the original light source with no inkling as to what was inside the cloaked region. Variations on this idea have also been proposed for seismic waves, which might one day be used to deflect an earthquake around an entire building or city.

Opposite: A network of osteocytes trapped in bone material; the brain of your skeleton. These cells are bone building cells that remain trapped in the bone matrix. They collectively make decisions about where to add and subtract bone by sending signals to the bone building and dismantling cells on the surface. If the bone breaks or fractures these are the cells that will signal to repair the damage, by emitting morphogens called growth factors to stimulate repair.

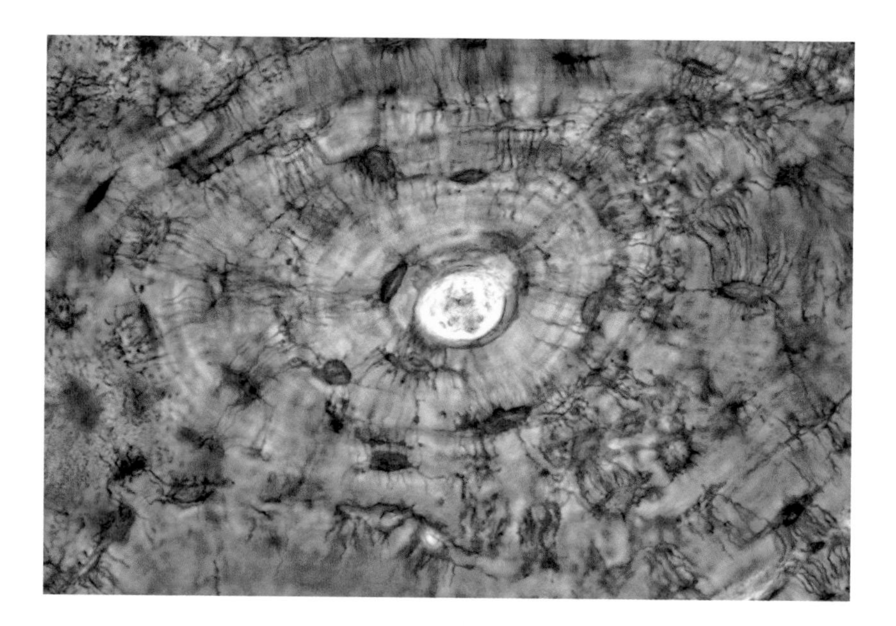

interpreted as looking at a system from every possible perspective. Such a notion arises naturally in quantum mechanics, where Bohm's expertise lies, because quantum systems can explore all their states simultaneously. It may not be possible to have an holistic perspective of a system if the system's set of possible configurations is vast. Fortunately, living organisms have found solutions for systematically searching vast spaces.

For example, the million-strong colonies of leafcutter ants that patrol tropical forest floors and the rafting fire ants that can survive a flood emerged because life built a system—evolution—that is optimized to explore many options. The lucky ant that finds a piece of food does so randomly—the ant is lucky, but the colony is not. The colony was guaranteed to find food because ants were searching the entire area. The lucky ant lays a trail of chemicals—known as a pheromone trail—to the colony, increasing the chance that more ants will find the food source and further reinforce the trail. Eventually a solid line emerges from a random set of meandering ants. By searching the whole space, optimal pathways between the ant nest and food sources emerge through reinforcement.

Another example comes from free-market economics, in which a market fully explores ways of trading resources to add value. Total factor productivity is a measure of how much emergent value has been added beside what you would expect for the inputs. You could argue the market is a place for discovering such emergent value. The entrepreneur who builds such a profitable enterprise is lucky, but the market is not.

These two systems have something in common: they have a finite framework. The search space is well defined and the parameters for hunting are clear. The search space explored by evolution using 4 nucleotides and 20 amino acids is similarly vast but *well defined and finite*. By restricting our own material searches to explore finite and well-defined frameworks, we massively increase the odds of finding an interesting solution.

The importance of history

In 2000 chemist Gavin Crooks made significant progress in non-equilibrium theory with the important observation that the pathway taken by a system as it evolves is, in some sense, a more fundamental concept than its state at a given time. This idea is exemplified in a simple chess game—whose course cannot be predicted from the rules alone. According to systems scientist Peter Corning: "*The 'system' involves more than the rules of the game. It also includes the players and their unfolding, moment-by-moment decisions among a very large number of available options at each choice point. The game of chess is inescapably historical.*"

Evolution too shows us the importance of history. Pandas evolved from bear-like carnivorous ancestors with fused fingers—a club with which to stun prey. So, when the ancestor of modern pandas switched to a diet of bamboo, their ability to adapt was limited by their evolutionary history. It wasn't so easy to evolve back to a five-fingered hand, instead the panda evolved a new makeshift thumb from a tiny wrist bone. But this elongated bone is hardly a substitute for a dexterous thumb, as anyone who has ever spent any time watching pandas eat would attest. Similarly, our own industrial history has set us on a trajectory toward climate change and pollution that is already rendering

Opposite: Imagine a system of microscopic rods that self-assembles under Brownian motion to form higher order structures. If the structures are able to fall apart again—as molecular structures can—the effective pathway taken by the system could fluctuate over time, enabling rapid exploration of a huge space of possibilities. From a given start point, the range of futures depends heavily on how easy it is for the pathway to fluctuate. This contrasts with macroscopic systems like chess or evolution, whose steps, once taken, cannot always be taken back.

Chess board pathways

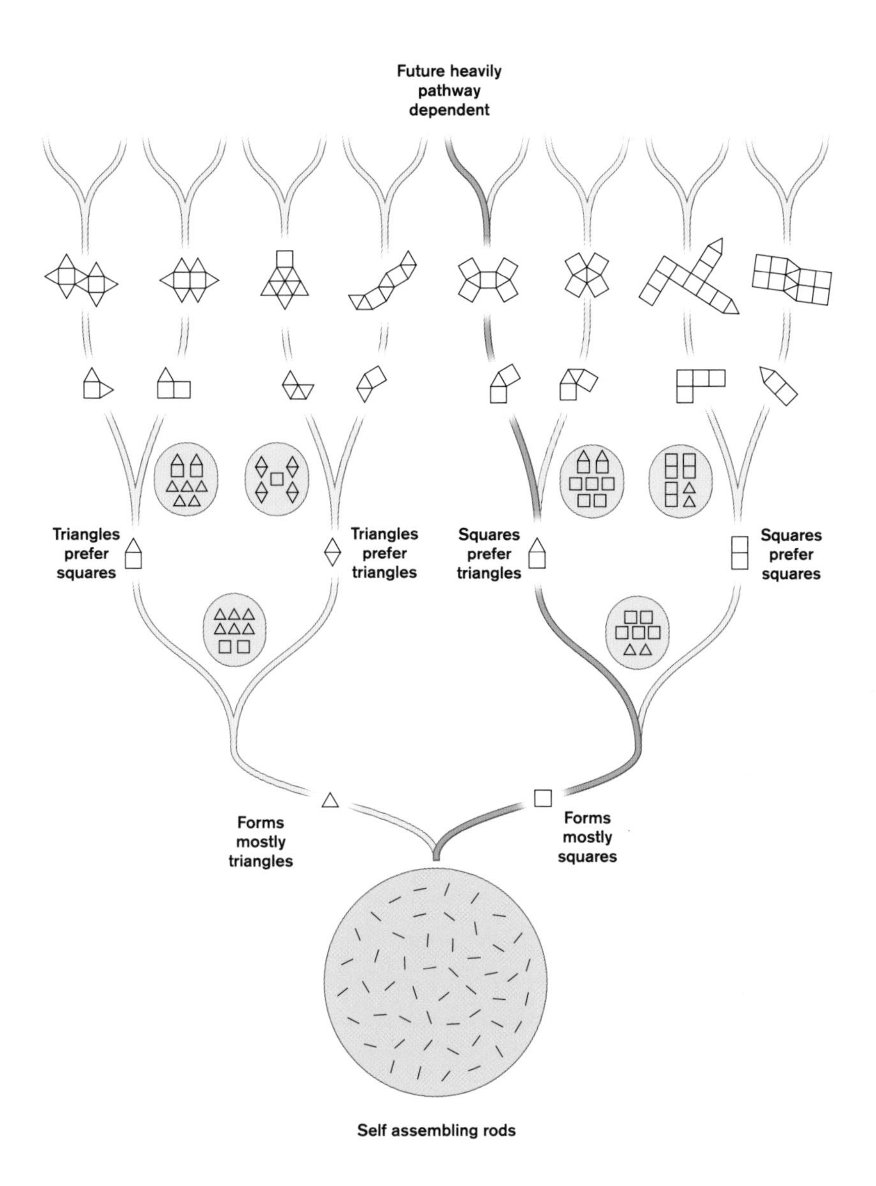

some regions of our planet uninhabitable. This linear system necessitates the build-up of waste materials and leaves us with aging technological infrastructure that was not designed to be repaired or regenerated; an indelible legacy, like the panda's thumb.

However, this economic model *has* created a broad set of existing technologies and industries that we can *choose* to bring forward into a circular economy. By choosing only those that can fit within the global biological and technical flows, and coupling them with emerging new technical infrastructures such as synthetic biology stacks and dissipative additive manufacture, we could start to construct our own industrial ecosystems (page 76). A network of supply chain pyramids, each with its roots in landfill and its branches intertwined with the very fabric of our homes.

Emerging Futures

A wise choice of technology for a circular economy framework could greatly simplify the conversion. The fewer materials we use the simpler it will be to implement, and the easier it will be to undo errors or retreat from unhelpful pathways as we explore the material space, by recycling them out of existence. Repair and regeneration of valuable infrastructure and products could also be simpler. The space of possibilities within the synthetic biology and AM paradigm is vast but it is finite. It creates a phenomenal playground in which human industry can operate. The simplified palette of molecules explored by evolution is rich enough to span the range of emergent behaviors that life on Earth has needed to survive for billions of years—including the ability to catalyze virtually any chemistry by folding proteins into almost any shape. In this chapter we have shown that joining together the frameworks of synthetic biology and additive manufacturing has the potential to support both halves of the circular economy, and this discussion has highlighted powerful ideas about the overlap between computation and chemistry.

Computation is the ideal tool to help synthetic biology make the leap from molecules to materials, and as we will see in the next chapter, in doing so artificial intelligence will prove essential in our transition toward material circularity.

Emergent phone casing

How could we build an emergent outer case for our biosmartphone? The specifications for an outer shell that can survive the rigors of the real world are demanding. We must integrate rigidity, toughness, functionality (such as a screen), and aesthetics.

Bone and wood exhibit many of the qualities we might want in our phone casing. A third choice might be chitin, which is a polymer made of sugar, used by many invertebrates in their shells and casings. For example, insect and crustacean exoskeletons and the mother of pearl that lines some mollusk shells are composed of an interwoven lattice of chitin and protein nanofibers—mineralized by calcium carbonate—that align in a hierarchical structure. Also, squid use chitin nanofibers embedded in a protein matrix to construct their beak—one of the stiffest organic materials on Earth. Within the rigid outer casing we could embed a bioinspired screen. Perhaps the most sophisticated biological example of a naturally occurring display screen is the dynamic camouflage found in cephalopods—squid, octopus, and cuttlefish—who can rapidly change their color, texture, and pattern to match their surroundings. There is great variability among species, but they generally rely on three types of specialized cells:

- **Chromatophore** cells are filled with brown, red, or yellow pigment, and can open and close like the iris in your eye. When open, they only admit certain colors.
- **Iridocyte** cells contain an iridescent organelle made from a stack of alternating layers of protein and extra-cellular medium which use the same principle of thin-film interference employed in microchip fabrication (page 62). The result is they reflect pink, green, yellow, or blue light, depending on the angle of view. Some squid can adjust the layer thickness to change color.
- **Leucophore** cells scatter light evenly in every direction, helping to create contrast against the other types of cell.

Such a device would be beautiful, with intricate lattice supports coated in iridescent mother of pearl. The completely transparent screen would reflect ambient light only where synthetic iridocytes were active.

7 Artificial Intelligence Algorithms

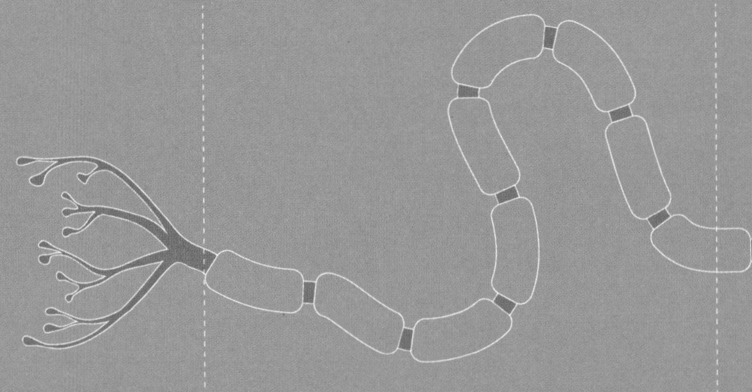

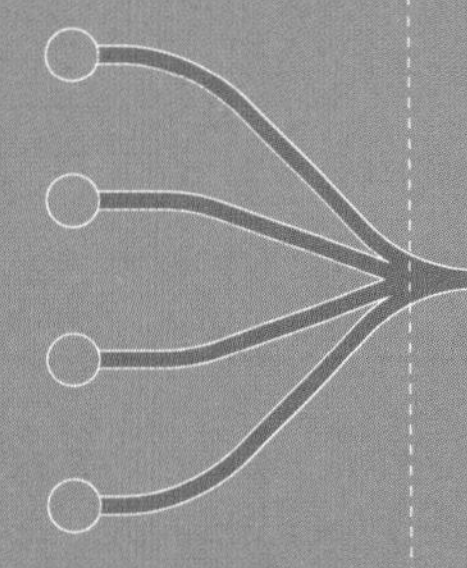

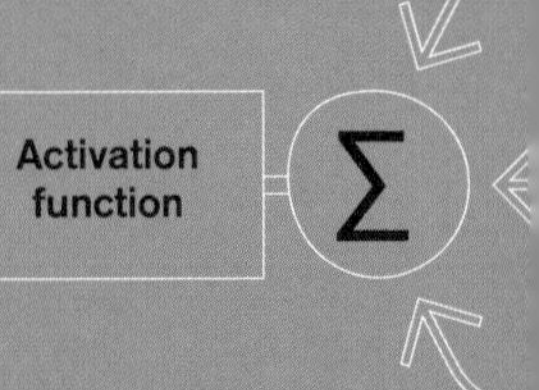

Natural and artificial neurons

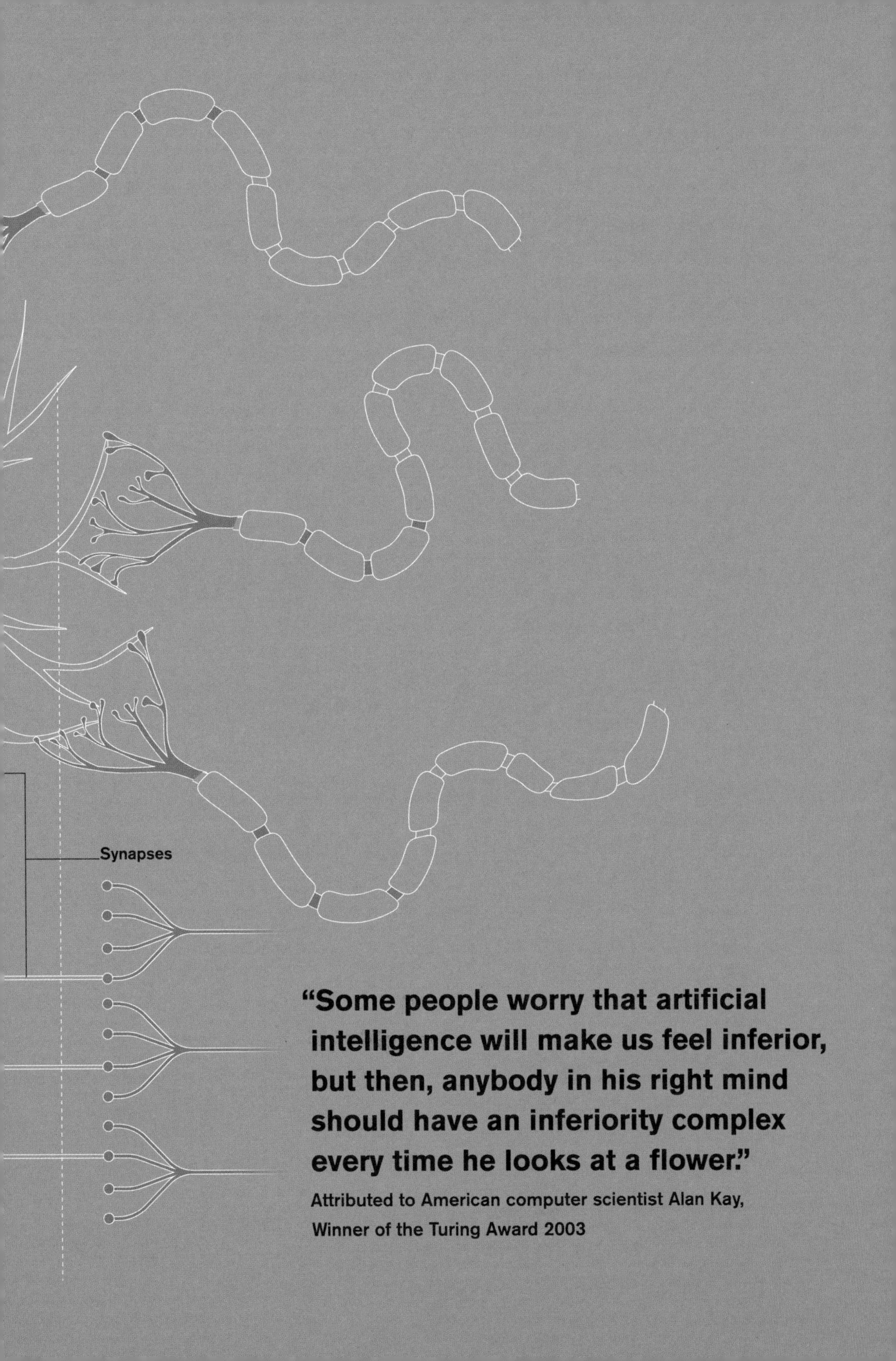

“Some people worry that artificial intelligence will make us feel inferior, but then, anybody in his right mind should have an inferiority complex every time he looks at a flower.”

Attributed to American computer scientist Alan Kay, Winner of the Turing Award 2003

The use of software to develop genetic circuits in synthetic biology and generate complex AM designs is well under way. Many designs, such as the evolved metal structures that resemble trabecular bone (page 86), have gone some way toward achieving the ideas about emergence presented in chapter 6, and we have developed 3D printing hardware capable of realizing such complex, static designs.

In considering the phenomenon of emergence, we realized that one approach to circularity was to carefully restrict the palette of chemistry that is used (making it easier for different industrial processes to exchange materials and waste), whilst maximizing the utility of these basic building blocks by exploiting useful emergent behaviors. But this is no small challenge—picking the right core chemicals, predicting emergent properties, and designing hierarchical materials represents a herculean task, even with continued advancements in computing. The game-changer will be artificial intelligence (AI).

With AI in our toolkit, we could vastly accelerate the process of selecting the right chemical building blocks and once chosen, AI will be essential for discovering and testing new materials, within that restricted (but still vast) chemical space, that can fit into an emerging circular economy. What is more, we can move beyond static materials and add an element of dynamism to our chemical spaces. Software that combines computer simulations and experimental data analysis with real-time monitoring of dynamically adaptive systems will enable 4D- and dissipative-manufacturing systems to emerge that can make use of the programmability of biological cells and the emergent behavior of materials in our everyday manufacturing.

The magic ingredient

The term AI is ubiquitous. While most of us have an intuitive sense of what it means, few would feel confident writing a formal definition. That's because AI is a general term, which applies to radically different systems in different contexts. AI refers to machines controlled by algorithms that display some degree of human-like ability to apply knowledge and skills. Artificially intelligent machines can sense and respond to their environment, solve problems and mathematical equations, or recognize patterns in data. Some forms of AI even learn and improve their own coding over time.

To understand how AI can help unite additive manufacturing and bioengineering to produce the smart materials discussed in previous chapters, it is helpful first to learn some fundamental concepts in this rapidly developing field.

Good old-fashioned AI

AI is divided into two classes: general and narrow. General AI is what most people think of when they hear the term—a computer or robot that displays human-level intelligence across all domains and tasks—but you're more likely to have encountered, and used, narrow AI. This class performs specific tasks at human or greater-than-human levels of skill, but cannot apply that skill to any other scenario. Narrow AI is all around us, from the virtual assistant in your smartphone to expert systems that help doctors analyze data and devise treatment plans. AI has transformed web analytics, stock trading, and surveillance, and is now commonplace in navigation systems and medical imaging, as well as many household objects.

The earliest forms of AI, colloquially known as Good Old-Fashioned Artificial Intelligence or GOFAI for short, were programmed like a piece of software—with problem, logic, and search functions represented by code—and remained the dominant approach until the late 1980s. GOFAI has been very successful in producing expert systems, which are able to interpret data to make decisions from a set of predefined outcomes.

Artificial brains

The fundamental method our brain uses to store and process information is the inspiration for a second branch of AI, known as machine learning (ML). Human intelligence could be described as emerging—somewhat mysteriously—from the interconnectivity and signaling between neurons in the brain. One of the most common types of ML—artificial neural networks (ANNs)—are nets of processing units, called nodes, which emulate certain aspects of the behavior of biological neurons.

Brains process information using complex networks of neurons connected by junctions called synapses. At these junctions, the information—initially in the form of a pulse of electrical energy—is converted into chemical messengers called neurotransmitters that travel across the void of space in the synapse and bind to receptors on the other side. A neuron will only send signals out when the combined strength of all its own input signals—often many hundreds of synapses—exceeds a threshold, called a weighting. When the brain learns, it changes the strength of that weighting—the total amount of neurotransmitters that must travel across the synapses—and therefore the likelihood that a receiving neuron will exceed its weighting and pass the signal on through the network. The combination of strengths across all the connections between all the neurons is the physical representation of learned information stored in the brain.

Nodes in ANNs mimic this information-processing behavior by signaling to other nodes only when their combined input exceeds their weighting. The result is that correlated pathways of signals (artificial thoughts?) can emerge in neural networks that connect an input pattern to the correct output pattern, much like a trail of ants emerging between food and nest. ANNs learn which weights to apply from a set of training data. Later, when they experience *novel* input, the correct pathways through the network fire to create an appropriate output in the unknown situation. As with human brains, the inner workings of the network—how the ANN arrived at the node weightings it has—are a black box, hidden from the programmer.

Unlike a step-by-step algorithm, the ANN coding contains no prior assumptions about the training data, and as a result the ANN "sees" the patterns and relationships

---

Opposite: An expert system (top) employs traditional programming to give users access to the benefits of a knowledge database compiled by human experts. In contrast, machine learning methods use artificial neural networks—inspired by the networks of neurons in the brain—to classify data. In supervised learning (center top), the strength of the connections in the network is set using training data where the key features have been manually labeled. Deep learning algorithms (center bottom) can learn unsupervised by using more complex multi-layered networks to identify the key features automatically. Some ML algorithms can self-improve (bottom) by feeding back information about the quality of the output into the weightings of the neural network.

Types of artificial intelligence

Traditional programming

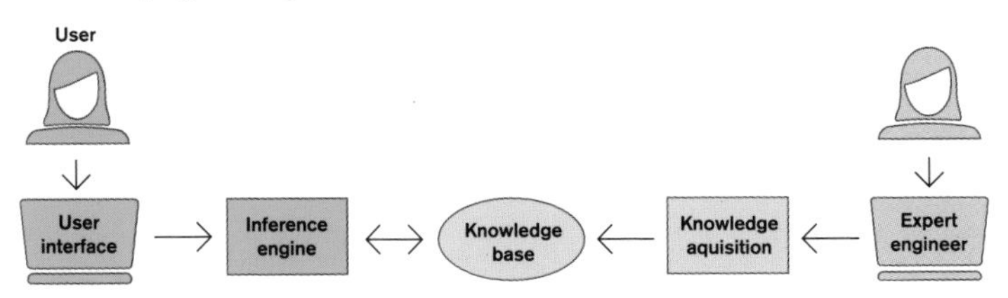

Supervised learning

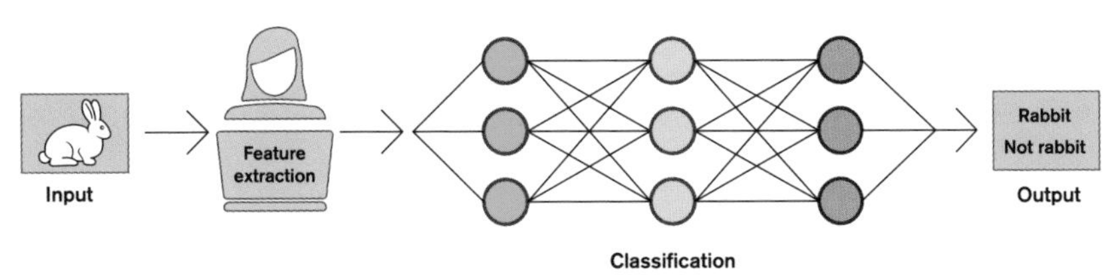

Unsupervised learning

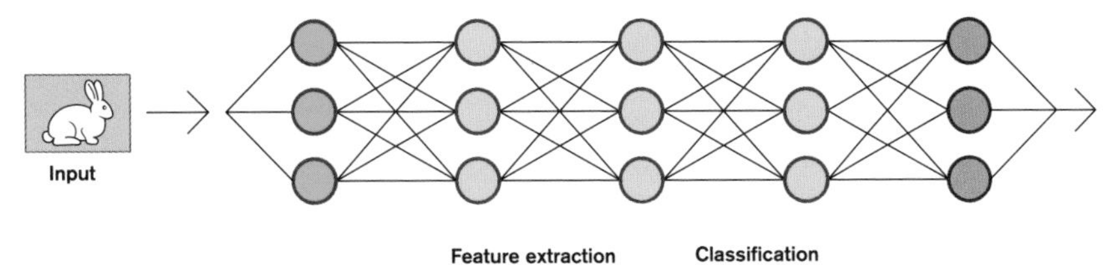

Reinforced learning

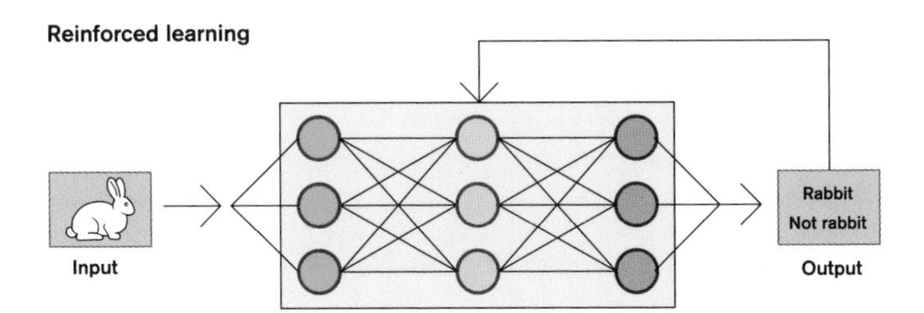

that might elude a programmer who uses traditional statistical analyses and may be influenced by emotions, expectations, prejudice or past experiences. This data-driven approach has allowed ANNs to be successfully applied in many fields, including medicine, agriculture, manufacturing, education, energy, transport, aviation, security, engineering, computing, banking, insurance, marketing, and art.

However, the collection of weights that ANNs develop is only as good as the training data—any errors, omissions or biases in the original data will be learned by the network, just as children learn values and behaviors from people around them. We must be mindful to train AIs on globally diverse and representative datasets, to avoid disadvantaging certain groups, or introducing other harmful biases. As the saying goes: "garbage in, garbage out."

Artificially evolved

Allowing ANNs to use evolutionary principles to optimize the pattern of connections between the nodes and the weights of those nodes—a technique known as neuroevolution—enables AIs to explore entire design spaces very efficiently. This is not so different from nature: you only have to look at a platypus to see clear evidence of evolution's penchant for lateral thinking.

A set of random ANNs are tested at a specific task, and the characteristics of the best-performing ANNs are combined and passed on—with a little random mutation—to the next generation. Essentially, neuroevolution is an algorithm for evolving ANNs and over successive generations their performance improves.

Although conceptually simple, neuroevolution yields remarkably complex results. For example, an algorithm called NEAT optimized the ANN used by the Tevatron particle accelerator at the Fermilab in Illinois to calculate the mass of sub-atomic particles called quarks, until it ceased operations in 2011. NEAT is particularly interesting because it tweaks the normal rules of natural selection by allowing new innovations a chance to optimize before the full force of selection is applied.

The neuronal weights passed between generations of ANNs are learned from experience, setting neuroevolution apart from biological evolution, where learned experiences predominantly do not affect the genetic information passed to the next generation. Thus, we can think of neuroevolution as more akin to human cultural evolution. Humanity's progress has been boosted enormously by culture, which allows society to accumulate knowledge collectively over time. Passing on such learned knowledge could help high-performing algorithms evolve much faster.

The challenge of general intelligence

Problem-solving, logic, and pattern recognition, although impressive, do not indicate understanding on the part of a computer. The narrow AIs explored here do not equate to human intelligence, which is based on more than just pure logic—it includes perception, intuition, emotion, spirituality, and experience.

In 1950 Turing defined a test that deems a machine to be intelligent if it can convince a human evaluator that it, too, is human. The engineers behind several AIs have now claimed their creations have passed the Turing test, some more convincingly than others. However, as this milestone has come and gone, we have realized it is not a true test of general intelligence. Other human cognitive abilities, such as common sense, understanding natural language, and real-world navigation, have been far more challenging to simulate with machine intelligence than pioneers in the field expected.

General AI may still elude us, but narrow AI will be essential in our *Brave Green World* to pull together everything we have discussed so far—additive fabrication using synthetic biological feedstocks to create emergent materials that fit into a network of manufacturing-, industrial-, and domestic-flows that self-replenish within a circular economy.

Automated exploration of material landscapes

The ability of artificial intelligence to find patterns in huge datasets, evaluate complex simulations, and make predictions about novel data or future conditions will make this technology an invaluable asset in our search for the right building blocks, materials, and emergent properties to form a circular economy.

But before we can start to explore the vast space of materials in front of us, we need a standard method of representing them and a way to search this library of standardized representations. Energy landscapes offer a useful framework to help scientists and engineers hunt for materials and are based on the internal energy barriers we have mentioned. The energy landscape for a particular molecule is a mathematical function constructed by computing the internal energy barriers between every possible arrangement of that molecule, and the final function resembles a mountain range. Some arrangements of the molecule have higher energy, which corresponds to mountain

AI supremacy

A legitimate concern exists: what if the machines became smarter than humans? A superintelligent being motivated to support and nurture humanity could lead us to utopia. A malevolent superintelligence could lead to our extinction, and perhaps that of all life on Earth, or an ambivalent AI might destroy us incidentally on the path to enacting its own goals. These concerns are not to be taken lightly. Experts generally agree there is high probability that human-level artificial general intelligence will arrive within the century, but they are divided over the risk it poses. Some have warned that self-improving or evolving algorithms embodied within an artificial general intelligence could become smarter than the smartest humans, in the blink of an eye. Whether these new forms of consciousness act favorably or harshly to humans is likely to come down to whether our goals align and whether we are in competition over limited resources. If we are able to master the principles of a circular economy, resources may cease to be a source of conflict. Nevertheless, researchers should carefully consider how they program intelligent machines, and governments might investigate regulatory steps that could reduce the chances of a malicious superintelligent AI emerging.

peaks, and other arrangements have a lower energy, which corresponds to the valleys. And the deepest valley—the lowest energy confirmation—is where the molecule can lose no more energy via earthshine! If it gets there and there's no new input energy, it will hang out there indefinitely.

If we compute the energy landscape, we can know everything about a molecule's behavior. The size of the peaks between the valleys determines how hard it is to rearrange the molecule as it moves around the landscape. Enzymes and catalysts work by effectively reducing the height of certain peaks, allowing molecules involved in a reaction to explore the landscape more easily.

For example, a metal might have several deep valleys corresponding to different crystalline structures. To rearrange this metal, we would need to heat it up considerably—applying a lot of energy in the process—to help overcome the high barrier walls. For a molecule like a protein, on the other hand, the surface might be more like a range of hills with one deep valley. This deep valley corresponds to its folded state, with many small barriers on the way. Thus, the molecule is able to explore different conformations with less input energy, the defining characteristic of soft matter (page 20).

Several different statistical strategies already exist to explore the energy landscapes of molecules and determine their properties. For example, molecular dynamics follows the random walk of a simulated molecule through its energy landscape and basin-hopping systematically explores the landscape by simulating different transformations of the molecule. Many kinds of software now exist to help explore different kinds of materials at different scales and represent their properties in libraries of data and software—known as multiscale modeling, for which a Nobel prize was awarded to Martin Karplus, Michael Levitt, and Arieh Warshel in 2013.

We could imagine every material that could exist or has existed—every atomic configuration of every element and compound—and represent it as a gargantuan digital library of energy landscapes. Each floor and corridor of the virtual library would house landscapes relating to similar molecules—proteins here, DNA there, metals there. Each book in the seemingly endless library would contain the landscape for that material, itself a seemingly endless vista of undulating hills and valleys. Seemingly endless because, from the point of view of someone stood within that material design landscape, the possibilities seem infinite. But the point is that they are not endless, there is a finite number of possible materials that could ever exist, given the available elements and basic physical laws. Finite but extremely large is what computers excel at.

What artificial agent, equivalent to a foraging ant, would we need to search such an incomprehensibly enormous materials space? The study of this idea is called materialomics. Researchers are already beginning to connect material simulation engines to automated and computer-controlled experiments that measure and map out the properties of useful materials. For instance, a deep learning program called REACTIVE generates new metamaterial (page 142) designs and tests their optical properties 200 times faster than conventional software. New materials that are stronger, lighter, or more flexible than conventional materials like steel or concrete are now being designed by AI, including a lightweight metamaterial that can be compressed to six percent of its original size.

If AI algorithms could be found to explore the full library we imagined above, then they could generate and navigate it far faster than we could hope to build the materials and run experiments. It wouldn't be necessary to compute every material—just one

Above: Mountains make great analogies for the mathematical functions called energy landscapes that scientists use to explain the properties of materials. Just as a mountaineer needs energy to climb, so a molecule needs energy to overcome the barriers separating different arrangements. The pattern of barriers determines the behaviour of the material. A crystalline material such as quartz might possess a landscape like Hawaii (top), with high and low energy barriers separating well defined valleys. The rolling hills of Tuscany (bottom)—with miles of similar sized peaks—might belong to a non-crystalline material like glass. Unlike natural landscapes, which are 3D, energy landscapes have millions of dimensions.

entry per floor might be enough! Once the outline of the library was sketched it might be possible to infer properties of energy landscapes that are not yet computed, to propose candidate materials for experimental realization.

Programming emergence

The examples of ants, neurons, and ANNs have shown us that relatively simple rules can produce emergent intelligent behavior. Similarly, simple computer models may be valuable for finding emergence in our materials library. For example, cellular automata are mathematical models that can produce elaborate, even beautiful, patterns. A cellular automaton is made up of a grid of squares, called cells, in which cells are either "dead" or "alive." Over time, cells flip between the two states using statistical rules based on their neighbors' states. Mostly, these rules produce uninteresting outputs, such as grids full of living or dead cells that never change state, but some rules produce elaborate emergent results such as fractals, oscillating or moving patterns, and even patterns that self-replicate—all common phenomena in nature. These behaviors are not programmed into the system; they emerge spontaneously from the rules, as in the mathematical models of the murmurations of birds (page 140).

Aside from being great models for how to find emergent properties with simple rule sets, cellular automata are even more interesting, mathematically speaking, because it has been proved that they are Turing Complete. That means cellular automata can possess the characteristics of a universal Turing machine (page 141), and a sufficiently large cellular automaton could compute any algorithm that a Turing machine could. Even simple computer algorithms, such as cellular automata could help us to explore large search spaces of potential materials, genetic circuits, or smart materials.

But programming behavioral rules—natural or artificial—to produce desired behaviours is no mean feat, especially when the environment can be highly variable. One dramatic example is found in army ants, whose nomadic colonies can comprise hundreds of thousands of ants, which although virtually blind, are capable of transporting more than 3,000 items of prey every hour by following the pheromone trails of other colony members. Yet their colonies can get caught in so-called ant-mills, in which thousands of ants die. Every so often a single ant gets separated from the 66ft (20m) wide foraging procession and gets confused, inadvertently bumping into its own foraging trail and creating a loop that starts to trap more and more ants. Unable to break free from their simple rule set, the ants are doomed to continue their pointless march until they die of exhaustion. Their unusual nomadic lifestyle has put army ants on an evolutionary knife-edge—strict adherence to the rules is essential to prevent the colony disbanding; following the rules too closely makes this milling behaviour more likely.

AI algorithms too, can similarly fall foul when the signals they have been trained on behave erratically. In high-frequency trading, hundreds of thousands of AI algorithms make millions of microtransactions, profiting from millisecond changes in stock prices, using relatively simple rules. In May 2010 when one of the key market signals the AIs had been trained to respond to—liquidity—moved out of the parameters of their training data, it led to rapid cycles of microtrading that sent the system spiraling further out of control. A built-in fail-safe temporarily shut down the stock market, breaking the cyclical pattern. The most divergent transactions were later nullified, reducing the consequences of this computational death spiral, although it could not halt the global recession that was already well underway.

Finding circular materials

Just like pandas, spiraling ants, or out-of-control stock markets, the slow rate of discovery of our current material exploration algorithm—largely trial and error—drives our economy into ruts. Once investments in certain materials—oil, steel, aluminum, concrete—have been made and the infrastructure for handling those materials—mines, refineries, mills and lathes—has been built, it becomes an expensive proposition to move away from assets frozen into those infrastructures. AIs could test millions of possible material configurations and production processes to find materials that can be manufactured within the global resource and energy constraints available. In this way

Opposite: Cellular automata are simple mathematical models that produce emergent results. Each cell in the grid has a state (represented here as different colors) and its state in the next generation (row) is determined by the current state of its neighbors. Rule 30 is extremely simple, yet produces surprisingly intricate patterns. Rule 104600 is more complex, using rules based on the average state of neighboring cells, so the final pattern depends heavily on the starting configuration. Some patterns naturally terminate (at steps 35 and 1517), while others may go on forever; termination cannot be predicted in advance. The results can be quite startling—one of the patterns shown seems to have taken on a vaguely human form.

Cellular automata

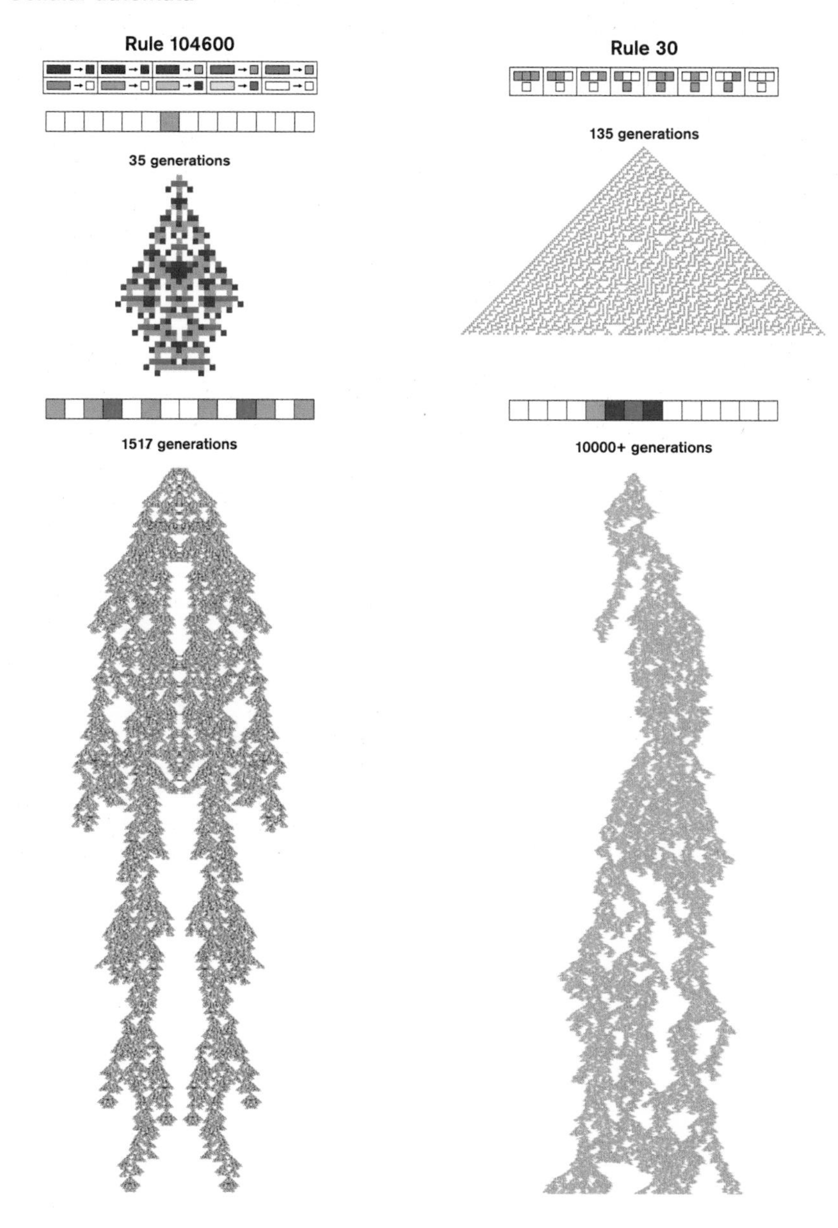

Above: Mathematical patterns such as fractals, seen here in a broccoli head (*Brassica oleracea*) (bottom right), and the Fibonacci sequence, seen here in a sunflower (*Helianthus*) (bottom left), are common in nature and can spontaneously emerge from some cellular automata rules. The triangular shell patterns of a sea snail called a textile cone (*Conus textile*) (top) are reminiscent of the alternating patterns produced by cellular automata Rule 30.

AI not only creates the material design landscape but acts as our compass to navigate through the plethora of new possibilities to find compatible sets of materials for a circular economy. If we started our search by exploring biological materials, we would likely find such sets much more quickly.

Despite occasional algorithmic meltdowns, ants are incredibly successful and part of the secret of their success is that they have behavioral plasticity—the ability to modify their behavior over time, for example, switching from random foraging to following a single trail. Their simple, environment-responsive rules allow them to adapt quickly to changing situations. In the evolution of human cultural rules, as with the inherited algorithmic behavior of our AIs, we must be wary of becoming locked into pathways that worked once but may not work again. Similarly, ideas that didn't work in the past may work in the future under different circumstances. We must choose a search space carefully and set up mechanisms for moving between appropriate solutions that fit the times.

Intelligent design and manufacture

AI's talents for simulation and problem-solving could be applied to designing products and hierarchical manufacturing sequences that take advantage of emergence. For a given purpose and set of parameters—such as materials, budget, manufacturing method—generative design software (page 87) could produce a wide range of designs, which can be chosen (or evolved) using machine learning. For example, Australian researchers artificially evolved the design for robot legs adapted to walk on particular terrains, like sand or rock.

AI-integrated 3D CAD software can be used to produce complex parts with improved geometric accuracy and consistency. The complexity of designing 4D objects, often combining multiple materials, is also well suited to the talents of such software. The software for calculating the G-code that gives the printer layer-by-layer instructions to build the object (page 85) is necessarily far more complex for 4D-printed designs, requiring intelligent CAD software able to simulate real-world scenarios and predict interactions between materials' internal geometry.

AIs can also produce virtual representations of highly-automated production plants known as digital twins. These virtual copies can simulate real-world factories in unprecedented detail, allowing virtual testing of new designs, materials, or procedures.

Indeed, digital twins can also couple to real factories via a network of sensors and actuators, enabling direct control or passive monitoring of the factory. Digital twins could predict deviations from the production plan, erratic machine behavior, or abnormalities in raw materials and assembled products. Indeed, unintended consequences such as faults are emergent properties; perhaps digital twins could also be used to explore emergent effects within the factory that are useful, rather than counterproductive. They could even be applied to factories processing synthetic biology crops, so that the organisms' DNA on a farm could be co-optimized in parallel with the factory for processing them.

Digital twinning doesn't have to stop at the factory walls. The impact of the factory in the local circular economy could also be understood virtually. Big data, analyzed by AI, could allow individual manufacturers to dynamically manage entire supply chains, evaluate and mitigate risks, predict customer behavior and identify sales targets, predict recall issues, design products, and more. Networks of such companies could work together to detect when supplies of materials are running low and find ways of re-tasking the industrial ecosystem (page 76) to optimize overall production. Companies might negotiate a price for changing the balance of industrial processes within one factory to produce more of the necessary ingredients for another. Indeed, the smaller and more local the circular economy, the faster the feedback and detection of emergence and the more feasible this entire idea becomes, which may be why biological organisms compartmentalize their "factories" into cells.

Sensors allow the detection of manufacturing defects so small no human expert could spot them. A wide range of sensors can be brought online for AI to make use of: infrared and ultraviolet sensors, ultrasound, ground-penetrating radar, or 3D laser-scanners, could all add additional sensory capacity to future AIs. For example, titanium manufacturers can use ultrasound sensors and torque monitors to perform predictive maintenance, listening out for wear and tear on the diamond tips of their cutting instruments and scheduling machines for maintenance before they break without warning.

Enhanced sensory capabilities also allow for precise remote control. For instance, London-based robotics company Ai Build has a suite of large-scale intelligent 3D printers including AiMaker, which can be attached to an industrial robotic arm and monitored remotely from anywhere in the world, using sensors and cameras to allow precise control over the chemistry and temperature of the printed material and early

Above: The intricate and unexpected patterns produced by some rule sets have singled out cellular automata as a potential source of inspiration in the design space. Cellular automata have been used in architecture—from creating artistic building façades (Cambridge North train station, bottom left) to innovative designs for future tower blocks (high-density buildings, top), as well as art and fashion (Pied de poule blouse, bottom right).

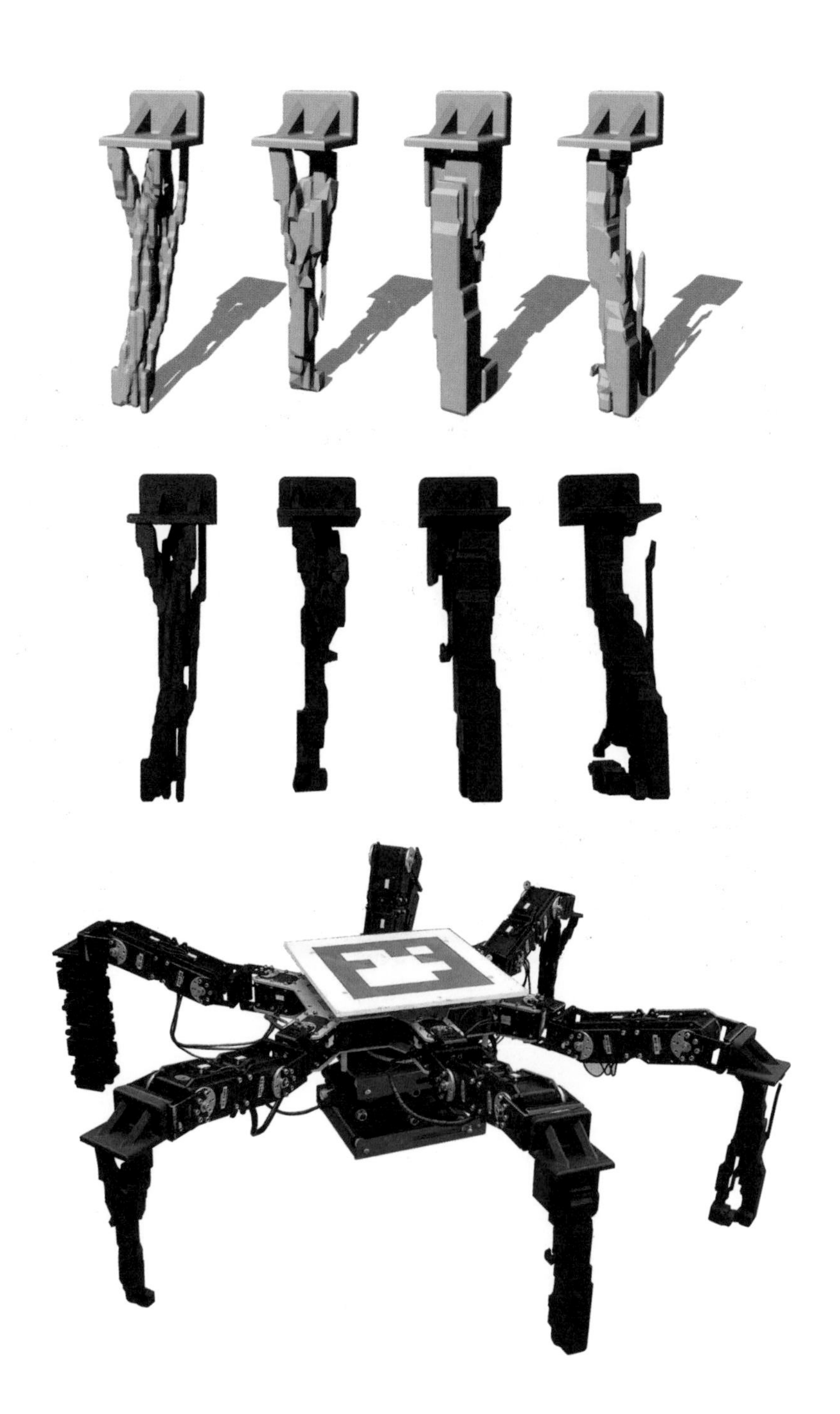

detection of defects and problems. Such large-scale 3D printing is poised to revolutionize construction. Engineers at MIT created an autonomous robot on motorized tracks to print entire buildings. The Digital Construction Platform (DCP) is intended for use in disaster zones, remote environments, and perhaps, one day, on other planets. The DCP's nozzle extrudes materials like concrete and insulation foam, switching between materials midway through construction.

Embodied intelligence

Modular designs will make it easier to apply neuroevolutionary concepts to successive generations of robotic hardware, such as those used in automated manufacture, as well as many other fields. As a proof of concept, Dutch professor and AI expert, Ágoston Endre Eiben, created EvoSphere, an environment with all the necessary conditions for robots to evolve (page 174). Here, 3D-printed modular robots, controlled using ANNs, are put through a learning phase followed by testing. Those with the best performance are selected to reproduce and create the next generation of 3D-printed robots. Combining this kind of approach with the AI-designed metamaterials we discussed earlier could lead to extraordinarily fast progress both in robotic design and product design more generally.

One way that some researchers hope to make use of AI in the real world is to control swarms of micro-scale or nanoscale robots. At these tiny scales, the reality of the machines we are discussing no longer represents robots of popular culture; these robots are actually cages of silicon, spheres of gold, or particles of nickel or iron, moved perhaps by magnetism or ultrasound. But why would we waste time designing nanoscale robots when they already exist in nature? Enzymes are nanoscale robots that can be programmed to perform tasks—and they have been programmed with DNA for billions of years.

Opposite: Jack Collins and colleagues at Queensland University of Technology in Brisbane used artificial evolution to optimize leg designs for a hexapod robot (bottom) by generating a variety of 3D models and testing their performance in simulated terrains. They mixed the characteristics of the best designs to produce a new leg shape. After 100 generations, spanning just ten days, optimized designs for walking through gravel, water, and soil (top), were ready to 3D print (center) and test in the real world. The legs met or surpassed the performance of human-designed equivalents. © Commonwealth Scientific and Industrial Research Organisation, 2019.

Above: Ai Build's robot-mounted 3D printer, AiMaker, has produced large yet delicate works of art like "Thallus," exhibited at Milan Design Week in 2017, designed by Zaha Hadid Architects. Combining fine-scale chemical control with a high-precision robotic arm allows large objects to be printed quickly and accurately.

AI meets synthetic biology

The powerful synthetic biology tools we explored in chapter 5 offer a glimpse of a future in which manufacturing biological components could become as everyday as smelting iron or molding plastic. Just as we imagined the materials landscape earlier in this chapter, evolutionary biologists often talk about the genetic landscape, which represents every single possible genetic configuration that could exist. High points on the map represent organisms with the highest genetic fitness (the number of offspring they produce that survive to adulthood, page 43). And like the proteins stuck in deep energy valleys, geneticists sometimes describe evolution as getting stuck on adaptive peaks or islands. If the series of evolutionary steps required for a species to shift from one adaptation—say a specific way of moving—to another solution includes individuals that would be significantly less fit than their parents, that evolutionary path will never be followed. The species will remain trapped on its current fitness peak, even if a nearby peak in the landscape would have even higher fitness.

AI can help navigate those valleys and create synthetic organisms for industrial, agricultural, and medicinal purposes that reach the highest fitness peaks, helping us to maximize the resource efficiency of each biochemical step in the circular economy. For instance, US biotech company TeselaGen's software program EVOLVE allows users to create, train, and use machine-learning tools to design genetic elements for a wide variety of commercial uses. The program creates vast libraries of thousands of different combinations of synthetic biological components, along with detailed instructions to assemble the DNA, which can be fine-tuned using the results of experimental tests. Asimov is now integrating AI into its CELLO platform (page 122) to develop a stack platform for living cells. They hope to combine the existing genetic circuit design platform with genomic, proteomic, and transcriptomic sequencing (page 109), machine learning, and mechanistic modeling of real-world physics, to further speed up the design of new genetic circuits.

Bringing it all together

AI may pave the way to the fourth industrial revolution, where computing, biology, and manufacture come together harmoniously, which could prove to be the turning point in humanity's transition to a circular economy.

Part of AI's role in the new circular economy will be to tackle the enormous challenge of sorting and extracting raw materials from our composite waste, just as living

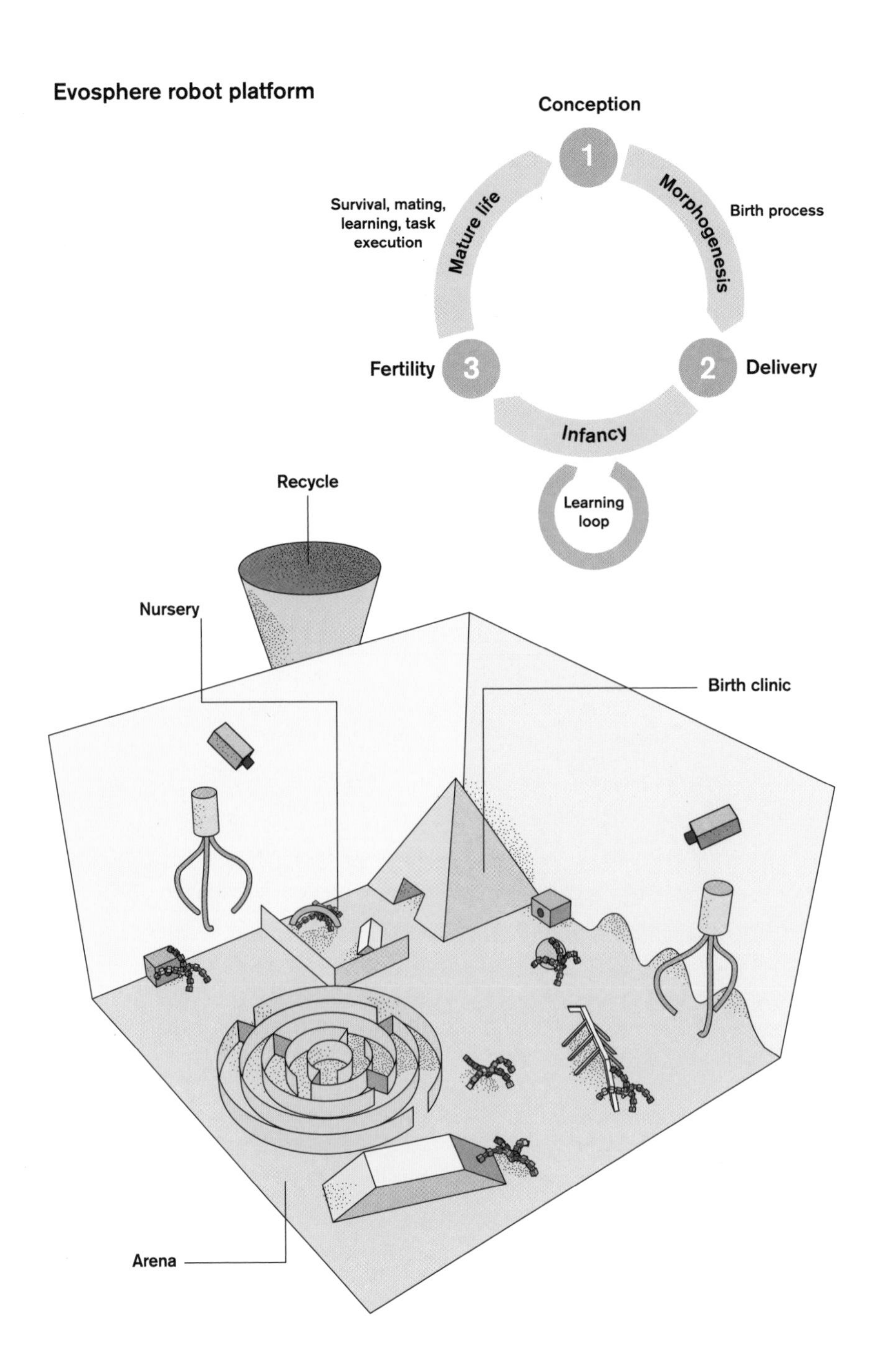
Evosphere robot platform
Conception
1
Morphogenesis
Birth process
2
Delivery
Infancy
Learning loop
3
Fertility
Mature life
Survival, mating, learning, task execution
Recycle
Nursery
Birth clinic
Arena

organisms can extract basic biological components from the air, water, and soil. AI will also help us live more efficiently: AI-integrated homes could monitor energy usage and take sensible actions to reduce consumption, and similar algorithms in industrial plants could control chemical reactions, monitor environmental conditions, and keep resource chains operating in sync.

Artificial intelligence algorithms will be the glue that binds our *Brave Green World*, uniting the basic physics and biology that we introduced in chapters 1 and 2 with the technological advancements we have reviewed in the last four chapters. From thermodynamics to genetic engineering, we are starting to see the threads converging on a circular vision for the future, as well as some tools for getting there. In the next chapter we will add the final piece of the puzzle—embedding computation in the very manufacturing process itself—that will allow us to take the technologies and principles we have discussed so far to the next level and start engineering our own industrial ecosystems.

Opposite: The Evosphere illustrates how robots constructed from modular, 3D printed components can undergo artificial evolution by mixing and matching their components to mimic sexual reproduction. Robots with random traits are fabricated in the birth clinic (step 1, conception) and move to the nursery where the AI can learn basic skills (step 2, delivery). Once trained, the robots enter the testing arena where their performance is measured and successful robots can "conceive" new robot offspring by combining their "genes" (step 3, fertility). Offspring inherit both the AI algorithm and their physical form, with a few simulated "mutations." With these parameters in place, artificial evolution can charge forward at full pace, producing robots that are better and better at their job with each generation.

Design and manufacture

Designing a device like a laptop or a smartphone is no simple task. It requires an iterative approach of testing and gradual improvement, and for our biosmartphone—in which the placement of every single molecule needs to be designed and optimized—the process becomes an order of magnitude more complex. It's one thing to iterate over the design and placement of a few thousand components, quite another to iterate over a trillion trillion atoms that come in 30 different kinds and can form hundreds of thousands of different molecules. So it is extremely likely that machine learning, combined with simulations and experiments, would need to be employed in the design of our biosmartphone and its fabrication—from the enzymes that build the chemical building blocks to the intricate hierarchical structures that form the outer casing.

To develop the basic building blocks, we might conduct experiments in microfluidic machines—already commonplace in laboratories around the world—where droplets of fluids, or synthetic biology cells, can be moved around using laser beams. Future fabrication chambers might be based on such fluidic technology in which droplets or synthetic cells seed the growth of a device by mineralizing a polymer gel to form synthetic bone. ML algorithms could direct mineralization in real time to fine tune the resulting hierarchical materials against design criteria. Once such a device is fabricated, either virtually or experimentally, artificial evolution could then select the best designs based on their performance in a range of scenarios. Selection pressure might be provided through feedback from real users printing devices at home, in a manner reminiscent of beta testing, where select customers trial early versions of new games and software before their public release.

With soft, flexible, and adaptive materials, our biosmartphone could take on a much more diverse range of shapes than a conventional smartphone, and AI-driven CAD could help develop new ergonomic designs. Perhaps such designs could be integrated within clothes, vehicles, or furniture, so that you didn't actually have to hold the device.

Fabrication time would be determined by the object's mass. To give us some idea, an apple takes about 100 to 200 days to grow, while a puppy takes 58 to 68 days to gestate, a human baby nine months, and an elephant 22 months. Plant growth is generally slower, but some species of bamboo can grow nearly a meter a day, fast enough to be noticeable to any human observers! So, a complete biosmartphone—assuming that it is roughly the same size as a modern smartphone—might take a few months to grow.

However, when comparing our device to a modern smartphone, we should consider the full lifecycle of the phone, from basic atomic building blocks in a mine, through fabrication and use, all the way to its degradation back into those basic building blocks again. Two months from atoms to phone is fast compared to the lifecycle for smartphone materials. Biological organisms begin aging and developing from day one, and a biosmartphone could also undergo a sort of ageing process at the end of its functional lifespan to prepare its sub-components for easy, automated recycling, in a matter of days..

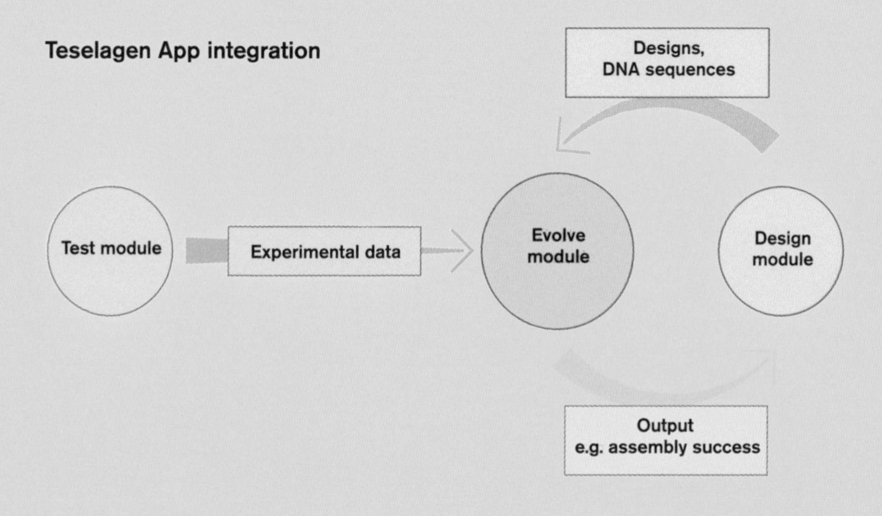

Above: The EVOLVE app works in collaboration with two other software modules: DESIGN and TEST, which provide DNA sequence designs and experimental validation data respectively to the EVOLVE module, which then designs the next iteration of libraries for testing. Over time, the experimental tests are used to train machine-learning models to assist scientists in refining their designs and assembly instructions more efficiently.

8 The Digital Process

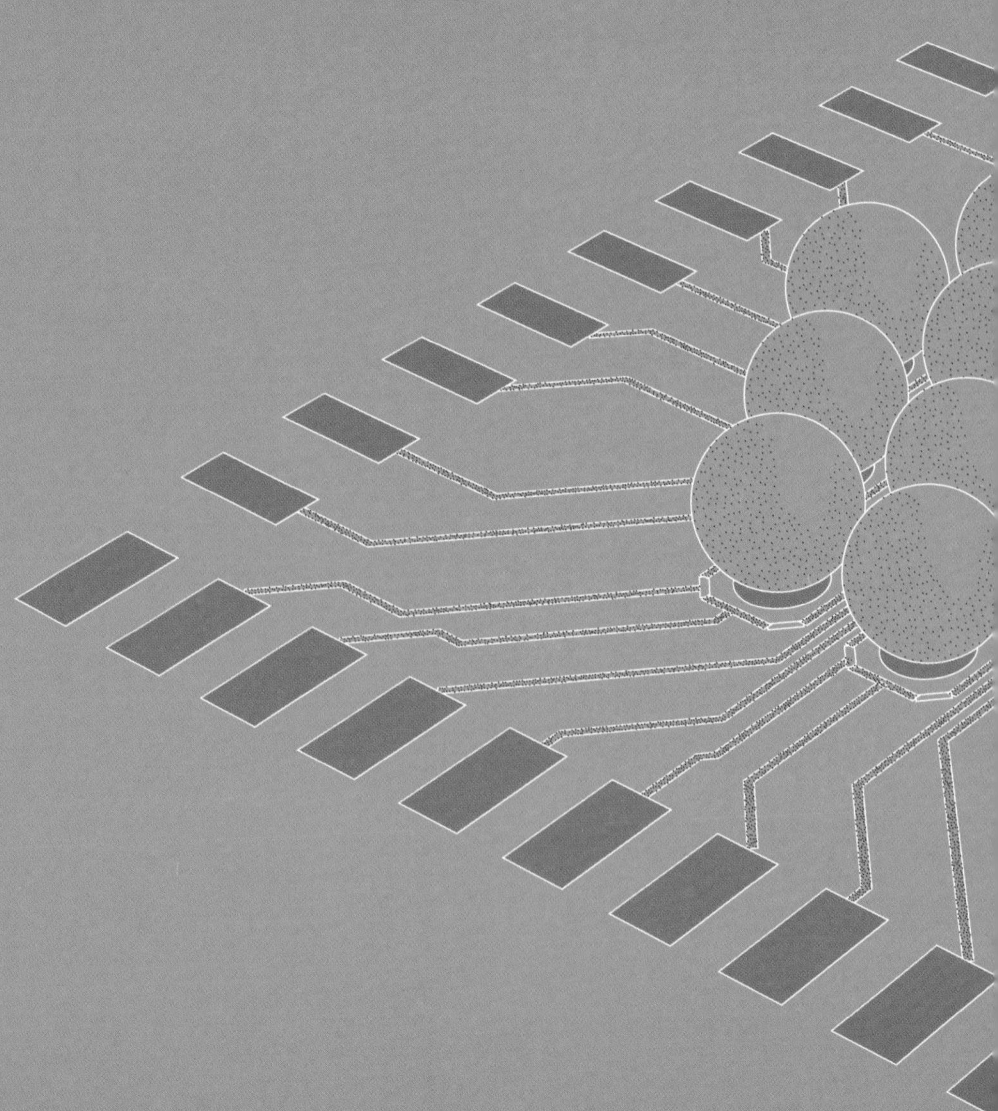

Droplet computer

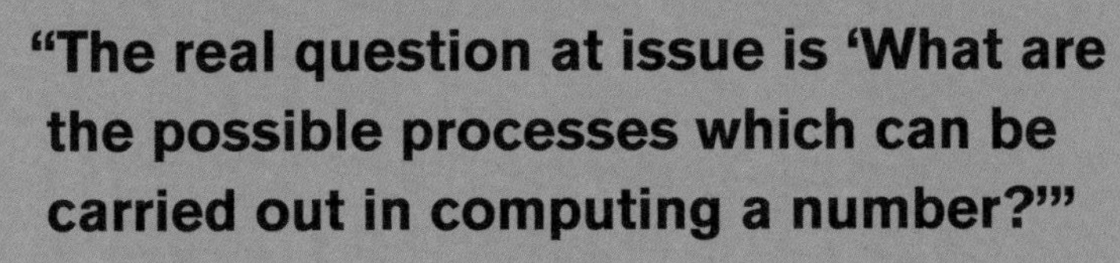

“The real question at issue is ‘What are the possible processes which can be carried out in computing a number?’”

Alan Turing, 1936

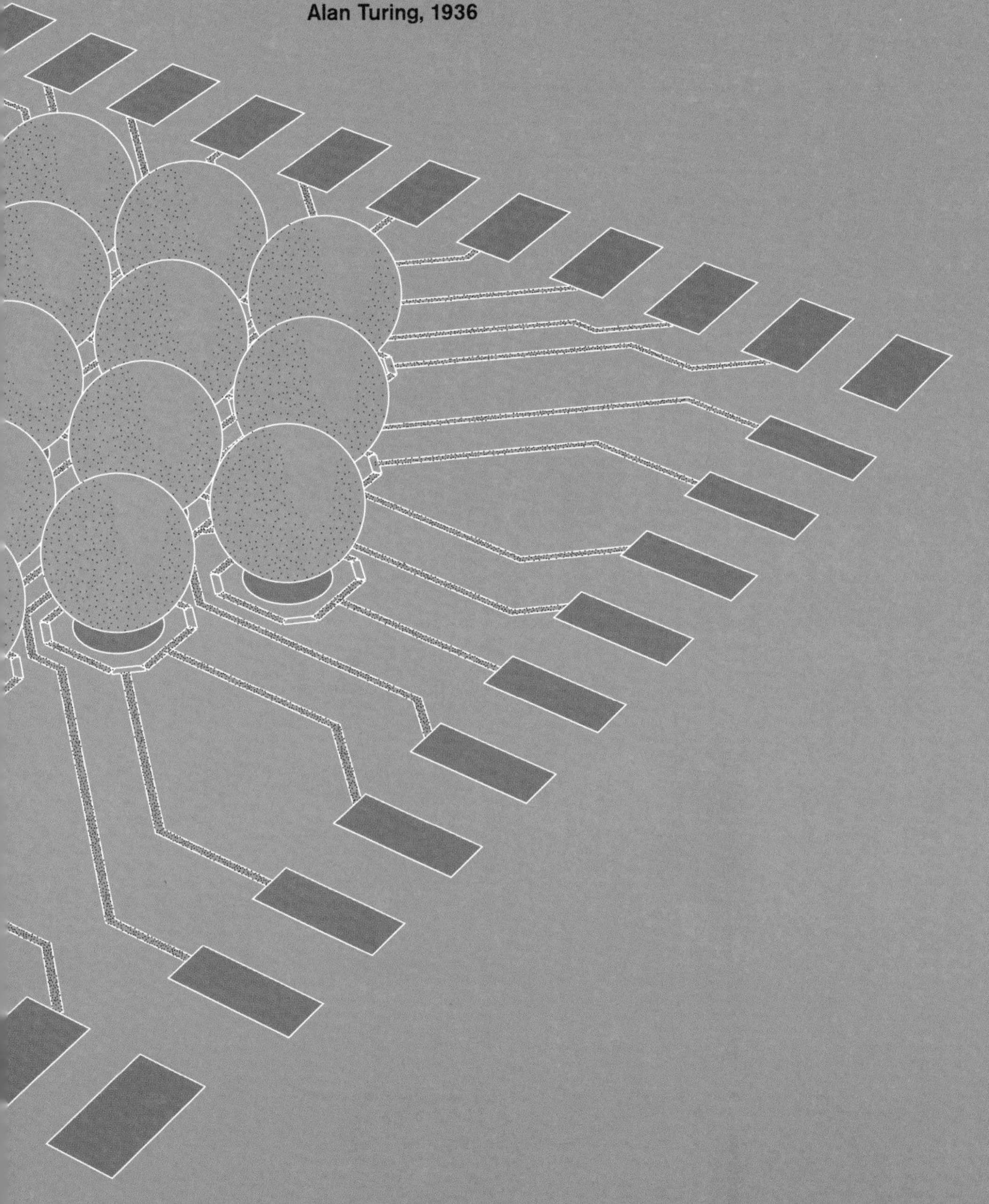

The possibilities that arise when advanced AIs develop algorithms for finding useful emergent properties are mind-blowing. Whether we are talking about chemicals, hierarchical materials, the design of genomes for plants, bacteria or fungi, or the digitally-twinned factories that process it all, the whole concept promises an advanced kind of material-centric technological ecosystem with non-living products and sub-systems that are at least as complex as living organisms. These ideas raise a question though: what kinds of products and systems will these incredible factories be able to manufacture in the future that we haven't thought of today?

In fact, this question can be turned on its head. Taking inspiration from the Alan Turing quote at the beginning of the chapter, instead of asking *"what can computers manufacture for us?"* we could instead ask *"what manufacturing processes can be carried out in computing a number?"*

Manufacturing as a form of computation

We could postulate the existence of a universal Turing machine (page 141), which assembled molecules, rather than writing symbols on a tape. *If* such a machine could be realized, then in the process of compiling molecules into a structure, it would simultaneously perform computation and fabrication in a manner vaguely reminiscent of a 3D printer.

If it worked, then anything that could be built from molecules with the manufacturing Turing machine—let's call it a "Turing mill"—would correspond to a computable number (page 141) in a vast mathematical space. Any product you could imagine, now or in the future, that was buildable with this molecular constructor would be indexed by a Gödel number (page 140) that encoded the process for making it—its program or algorithm. Assuming there was a suitable way of introducing the instructions, the universal Turing mill would be capable of making any object that could be expressed in this mathematical realm. The goose that lays the golden Gödels!

When a Turing machine runs an algorithm, it either finishes—returning an output—or it doesn't. If it finishes, then the output is known as a computable number; if the Turing mill was effective then it could build objects corresponding to computable numbers—a buildable object.

Since we know that cellular automata can implement Turing machines, we can take some of the ideas that apply to cellular automata and apply them to the Turing mill. For example, some cellular automata produce emergent patterns that aren't explicitly programmed into them. Perhaps then the Turing mill could produce objects that exhibited such emergent properties like these cellular automata. Indeed, some of the algorithms on the Turing mill would never finish computing, like a tree or fungus that never stops growing. These might be exactly the kind of algorithms needed to manage material flows in a circular economy, which also never end.

If these ideas could be formalized and realized, then in the same way that networks of computers combine to form the internet, our *entire* economy could become a giant distributed computer running an algorithm that never finishes, managing the technological flows of material in harmony with the biological flows, and all within the planetary boundaries. In that way, humans might survive on Earth for more than our 100-million-year goal, giving us time to spread out across the solar system and beyond.

Understanding molecules mathematically

The type of computational simulation our emergence-hunting AIs will need relies on numerical representations of the physical behavior of the atoms and molecules. To run those simulations on silicon processors the rules of chemistry and physics must be represented using arithmetic. That this can be done is remarkable, but it does mean that it takes a lot of computing power.

The results of any such molecular simulations are only as good as the approximations of real-world physics and chemistry built into their programming. If these numerical models of matter are wrong, or lacking in key features, then further computational analysis of their emergent behavior is a waste of time. Furthermore, in modeling extremely complex systems like organisms or hierarchical materials there are so many variables and unknowns that they may not surrender easily, or at all, to expression in a coherent mathematical or numerical way. The lack of a full mathematical formulation of non-equilibrium physics, or a generalized theory of ecological science, for example, could be major hurdles for designing 4D materials or synthetic biological dissipative systems to construct ecosystems.

To find beneficial emergent behaviors in simulations, we would need to run analyses on millions of atoms, which—with standard computing architectures—can take months, even for simple molecular systems for which the physical rules are well understood. The standard architecture of computers—designed to perform efficient mathematical operations to strings of information—may not be the kind of computer best suited for AIs to perform their material-hunting task. If we can systematically learn how to build computers in novel ways without silicon chips, we might be able to design different kinds of processors that are very well suited to solving specific types of problem, such as hunting through the energy landscapes of chemicals. Perhaps we could find one that is intrinsically good at simulating or exploring hierarchical materials.

A machine that could use self-assembly of materials or dissipation of energy as operations available to a computer calculation would have the physics and chemistry built in by default and might therefore be able to explore the space of materials with simpler programming than a traditional computer. Computers that use transistors to model basic atomic behavior, as well as the chemistry of their interactions—as current simulations of material properties must—could be expending vastly more computational effort than might be necessary if different forms of computing were available.

If we could formalize such a concept as the Turing mill at the molecular level, we could bring to bear all the results of computer science, and the centuries of mathematics that support it, onto a new form of direct digital manufacturing. The explosion of ease with which regular computers manipulate information could translate into manufacturing, enabling everyone to manipulate matter locally, maybe even as easily as changing the picture on a computer desktop.

When do physical systems compute?

A computer can be defined as any physical system that can perform an accurate mathematical calculation. The problem is knowing if any given physical system is a useful computer or not, and whether we could trust its output. Before running a computation, an abstract mathematical representation needs to be created of the problem—such as binary arithmetic—which is a process called embedding. Ideally this representation will readily translate into the physical medium of the computer. The mathematically embedded representation of the problem then needs to be portrayed in the physical hardware of the computer—such as the "on" and "off" transistors in a silicon chip—a process called encoding.

For example, if we wished to compute the share of a $150 dinner check between five people, we can embed the problem using logical symbols: 150/5. This has a well-formed meaning in our abstract mathematical representation, which in this case is simple arithmetic. Once we have a logical embedding, we now need to encode the problem into the physical realm. On a calculator we physically punch in the symbols: 150÷5= and the input data and algorithm to perform it are converted into a physical representation within the calculator by electronics that sense the key presses. Strings of transistors encode the numerals 150 and 5 as binary digits and the divide operation is selected as the algorithm to employ. The calculator implements a series of *physical* electronic processes that switch on or off various transistors, processing the input numerals. After the calculation, the physical representation of the output is converted into decimal and displayed on the screen as "30." Crucially, the physical processes in the transistors have a fully understood and realistic two-way relationship with the arithmetic they represent. We can say the relationship "commutes."

In principle, we do not have to use a calculator to perform the calculation. We could calculate it mentally or on paper. This allows us to test the relationship between the physical processing and the embedded arithmetic and make sure that the output of any

Mapping between mathematical and physical space

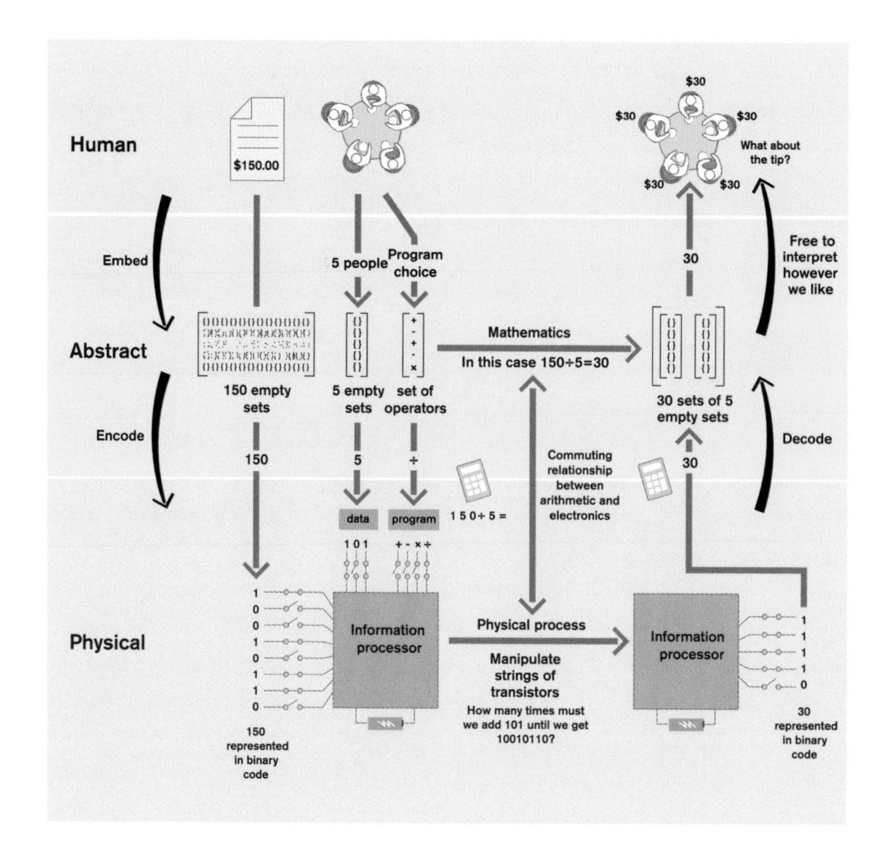

Above: A computational problem can be represented both mathematically and in a physical system. If both yield the same correct answer our physical system has performed a computation. In this example, five people split $150 evenly. First we *embed* the problem in arithmetic: 150 ÷ 5. Then we *encode* the problem on a calculator, which performs physical operations on strings of transistors—like a Turing machine rewriting symbols. Energized with a battery, the physics of the calculator computes our divide problem—like marbles falling through a Turing Tumble (see pages 186–187). We *decode* our final answer to get $30 each. The same machine can do this for any input number or algorithm that the machine has the capacity to represent.

computer matches the manual solution, building trust that the computer will give us an accurate result for calculations too complex for us to do by hand.

Ultimately, these mathematical ideas *might* extend to living organisms themselves. After all; biological growth is a combination of additive and subtractive manufacturing. We could speculate whether biological organisms are already a form of computing system, but in that case, the question becomes: what are they computing?

Computing in biological systems

An alternate definition of computing, postulated by Turing and American mathematician Alonzo Church, states that any physical process that can emulate a Turing machine is performing computation. Several existing biological processes resemble facets of Turing machines and implement physical processes that might be repurposed to build the Turing mill.

The ribosomes discussed in chapter 4 superficially seem to resemble Turing's machine. There's a mechanism for sensing symbols on one tape and an ability to write new symbols onto a second tape. But it cannot delete symbols or reverse the direction that the ribosome moves along the RNA strand.

Certain enzymes perform read-and-write operations on DNA, such as Taq polymerase and CRISPR-Cas9 (chapter 5). The CRISPR-Cas9 complex captures DNA molecules and incorporates their information into a second DNA molecule. Thus, CRISPR can also do many of the operations that define a Turing machine, but the enzyme is not "universal" because it only performs a single algorithm.

Bone-forming cells sense, add to, and subtract from skeletal material. Like a computer, bone fabrication uses external inputs (such as mechanical stress or availability of raw materials) to optimize a response—in this case maximizing bone strengthening while minimizing resource use. If a meaningful mathematical framework were assigned to bone construction, perhaps the system could be subverted, say to perform optimization computations, with the output encoded as a particular bone shape.

The fact that ribosomes, CRISPR, and bone growth resemble some aspects of Turing machines, or other models of computation, does *not* mean that organisms are computing. Not all arrangements of transistors perform computations! However, it is fair to say that organisms may be doing things that we could copy or co-opt to implement our machine.

Types of computer

In principle we could build a Turing mill—Turing and Gödel provided the theory—but theory is a world apart from building a practical computer of any sort. In the following series of examples, we gain insight into how other types of computers are built, which might give us some clues.

Turing Tumble

The Turing Tumble is a brilliant toy—for "children" of all ages—that explains the concepts of computing and programming well. We have a supply of red and blue marbles positioned at the top of our device. These marbles represent a supply of input data, like 0s and 1s in a binary code. The objective is to set up pathways through a sequence of components so that, under the action of gravity, the red and blue marbles arrive at the output channel in a specific pattern. If the colored marbles represent the supply of symbols available for a Turing machine to write, then the pattern at the end is the number that it is computing, and the arrangement of components is the algorithm.

Trainyard

A similar game is available in virtual form as an app called Trainyard, which starts with a series of colored locomotives in their respective sheds. The user must construct a system of tracks to link the trains to their final resting place in the destination trainshed, which only accepts trains of a certain color. Along the way trains can split, merge, be repainted or diverted down branches. The range of instructions is broad. Whatever happens, the right number of colored trains must end up at the specified location.

By comparing Trainyard and Turing Tumble we see that the nature of the instruction set—the operations that one can perform on the input symbols—is critical in defining the properties of the computer and how efficient it is at various kinds of calculations. In Trainyard you can repaint the trains; the Turing Tumble cannot repaint the marbles.

The Von Neumann architecture

Most modern computers use the Von Neumann architecture—based on a description written in 1945 by John von Neumann—which stores both the program and the data it runs on in the same memory system, so that the instructions are supplied to the processor alongside the data. The main difference with a Turing machine is that the Von Neumann

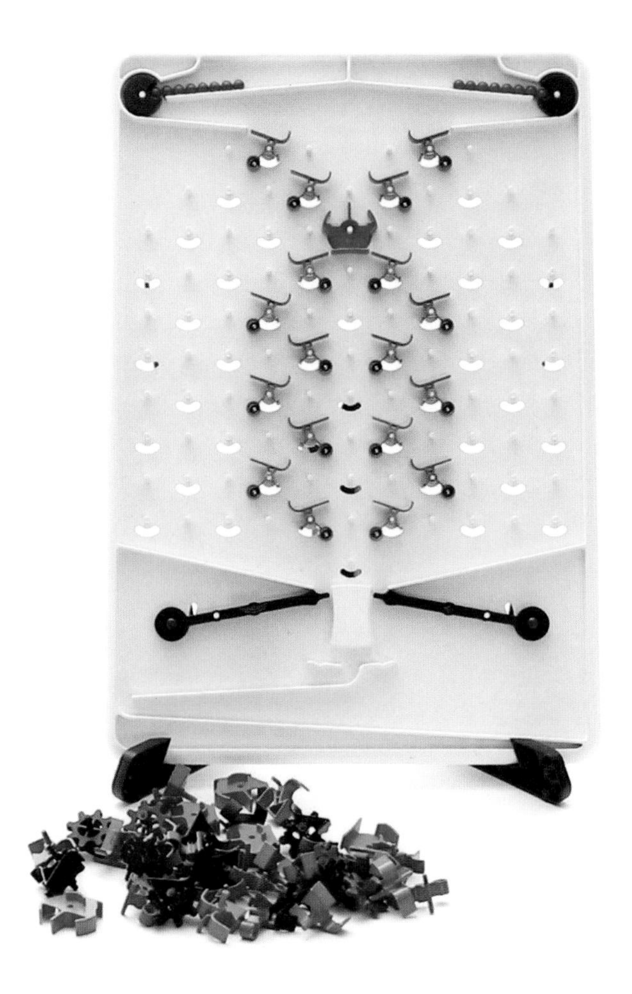

Above: By carefully setting up the components of the Turing Tumble that connect the input to the output channels, distinct pathways for the red and blue marbles can be set up, generating a wide variety of red and blue patterns that correspond to binary numbers.

memory is random access, allowing processors to access memory addresses without scrolling through others first—a major advantage over architectures that scroll along tape.

Like Turing Tumble and Trainyard, the processor takes input—here in the form of high- and low-voltage signals representing a sequence of 1s and 0s—and performs logical operations on the data with its arithmetic and logic unit. Such operations are analogous to the merge and split functions in Trainyard and essentially perform arithmetic on the data such as add, subtract, multiply, and divide.

Above: John Von Neumann, pictured here with an early computer at the Institute for Advanced Study, Princeton, USA in 1952, succeeded in the unenviable task of creating a practical computer architecture from his own abstract works and those of Turing and Gödel *et al.*

Quantum computers

A Turing machine that was also a quantum computer would be able to read and write quantum states into the Turing tape. In practice, this means the Turing machine could read and write an entire set of superimposed symbols—known as a qubit—into the same box on the tape, performing an operation on all the symbols at the same time. It would be as if the marbles in the Turing Tumble could be different amounts of red and blue *simultaneously*. By managing each qubit through the different pathways of the algorithm, different qubits could mix some of their red and blue with other qubits to become "entangled." What happened to one would affect the other. Different operations would result in different levels of mixing between red and blue. In the final output you would get a line of qubits all of which are entangled—at the quantum scale—with one another, analogous to the line of red/blue marbles at the bottom of the Turing Tumble.

Evolvable hardware

A beautiful example that isn't necessarily a computer, but could be, establishes a route for programming matter directly. The example relies on a device called a Field Programmable Gate Array (FPGA) and its key feature is that its hardware can be reconfigured using software. In 1996, Adrian Thompson, a researcher at the University of Sussex in the UK, reprogrammed an FPGA by evolving the string of code needed to program the FPGA so that it would perform the operation he was trying to implement.

When he analyzed the evolved circuit, it was nothing like the circuit that an engineer would devise. Our world of knowledge is based on abstractions, short cuts, and models rather than a full, true understanding of reality. Programming with the evolutionary technique gave access to the full physics of the device—whether we know that physics or not! Indeed, one part of the circuit was not even logically or electronically connected to its neighbors, but if removed, the device didn't function. Perhaps it was acting as a capacitor, which could have a filtering effect on the input signal?

Shining light through materials

A novel form of computation, explored by Prof Tim Wilkinson at Cambridge University, in the UK, is based on the way light behaves when it passes through objects. We have a good mathematical representation of the behavior of light in systems of lenses and optics, known as Fourier analysis, which sets up a nice commuting relationship and is

used in a variety of situations such as signal processing or image analysis. Perhaps it would be possible to use systems of lenses or photonic structures to perform Fourier analysis for us instead of computing them using a Von Neumann machine? It is possible to use generative design to evolve photonic structures, which can interact with a light signal to output complex interference patterns that depend on the nature of the input light. Such structures—similar to those found in the wings of birds and butterflies, where they generate vivid colors, as well as in thin films of oil on water, or the optical stacks used to make microchips—could allow us to process input light signals and perform extremely complex calculations in the blink of an eye. Once this "algorithm" had been hardcoded in a block of matter, modifying the input light would get different answers at the far end.

In-materio computation

Our foray into different kinds of computers taught us something valuable. The FPGA and light-computing examples demonstrate that there may be "natural computers" in the real world; arrangements of matter that respond to energy input in a way that

Opposite: A quantum computer makes use of quantum-mechanical phenomena, such as superposition, tunneling, and entanglement, to perform algorithms. One kind of mathematical problem at which they excel is the exploration of cost functions, such as energy landscapes. Quantum systems can tunnel through energy barriers allowing rapid exploration of a cost function. A quantum Turing machine could write a stack of symbols in a single place on the tape—as if a typewriter could write five letters at once and then choose later which of the five letters actually appeared. A symbol might depend on its neighbors further up or down the tape according to the rules of quantum entanglement.

Above: A Field Programmable Gate Array consists of an array of logic blocks with inputs and outputs that can be interconnected with their nearest neighbors to create complex logical operations in the hardware. The interconnects can be reconfigured even after the device is deployed, making it programmable in the field, hence its name.

humans might recognize as performing a useful mathematical function. They may not be Turing complete (page 163), but they might be good at specific problem-solving—rather like a narrow AI (page 155) or a non-universal Turing machine. This whole field of enquiry is called in-materio computation.

There may be more than one way to achieve computation in matter. We could, for example, build a chemical Turing machine solely to do computations rather than having any intention to do fabrication. Such a machine is a chemical computer, or chemputer.

How can we build a chemputer?

Three things are needed to build a chemputer: A mathematical representation in which we can embed an arbitrary problem, a way to encode the logical formulation into the chemicals, and a way to read out the result.

Let's assume that the chemputer will be made of molecules dissolved in some kind of liquid, so all the molecules are able to move around freely via Brownian motion. Typically, there might be a trillion trillion molecules in a tablespoon of solution. As these molecules diffuse, they encounter other molecules and react to form new ones. There are a staggering number of combinations of molecular configurations that could occur.

In fact, imagine a long line of droplets that plays the role of the tape in a Turing machine. Each droplet contains a chemical solution that can undergo internal chemical reactions. A network of possible molecular reactions can be sketched out based on the ingredients put into the chemical droplet. Initially, let's assume the number of atoms in each droplet never changes, but they will react with each other to form different configurations—different molecules. The relative amount of each type of molecule is determined by the energy barriers that govern the rates of reaction between the molecules currently in the network, and the amount of energy available to the system.

Let us further assume that each *type* of molecule in our droplet corresponds to the symbols that the chemputer would be able to read and write. Thus, at any one time, each droplet would have a certain amount of each "symbol" equal to the number of that kind of molecule currently in the droplet. In this picture we see a similarity with the quantum Turing machine, because we can store and manipulate multiple symbols in each droplet; the chemputer can handle mixtures of symbols too, although our chemical droplet symbols obey different rules than their quantum counterparts—our droplets probably couldn't become entangled in the quantum sense.

The molecules in each droplet can exchange individual atoms and molecular subgroups, with chemical bonds being created and destroyed in the process, which is analogous to marbles moving through a Turing Tumble or locomotives in Trainyard. Each operation available to the chemputer for reading and writing symbols—its instruction set—would involve driving statistically large numbers of atoms from one kind of molecular configuration to another through various pathways in the internal energy landscape of the droplet. In the process of such reactions entire sets of symbols would inter-convert.

A theoretical chemputer would sequentially look at a line of chemical droplets and measure the state via a readout mechanism, such as fluorescence or electrochemical potential to assess the amount of each molecular "symbol." Following its encoded algorithm, the chemputer would then write new states into each droplet, which could be achieved either by driving energy into the droplet to push the reaction network into a different state, by introducing or removing molecules to alter the energy landscape of the droplet, or by somehow directly modifying the internal energy barriers that govern the rates of reaction. The latter could be arrived at by adding or switching on catalysts or reaction-specific enzymes within the droplet using external triggers such as light signals.

As we set up the energy barriers that govern the reaction rates inside each droplet, we are establishing a maze through the hyperdimensional energy landscape, which our trillion trillion atoms will randomly explore very quickly. We can drive the random exploration in our chemical potential energy maze with an external driving force that is applied across the system, to ensure that the computation proceeds in the right direction. This idea was inspired by Charles Bennett's superb description of a computer system driven by Brownian motion in 1984.

In fact, in a regular silicon computer something similar is established—a potential energy maze for individual electrons created by energizing or de-energizing transistors. But in a droplet computer each molecule is free to diffuse, making the maze vastly more interconnected than the transistors in a microchip. The system is free to move in as many directions as there are atoms in the droplet, which outnumbers the number of transistors in a microchip by a *factor* of ten trillion.

Every possible state of the droplet is a single point in our hyperdimensional maze, in which corridors that connect different molecular states can be opened or closed by manipulating either the chemical driving force, or by enhancing or reducing energy barriers. The computational algorithms are implemented by moving the barriers of the

Describing chemistry for computation

Amount of each element

Hydrogen n_1
Helium n_2
Lithium n_3
Berylium n_4
Boron n_5
Carbon n_6
...
Tennessine n_{117}
Oganesson n_{118}
Total: 6×10^{23}

Amount of each molecule

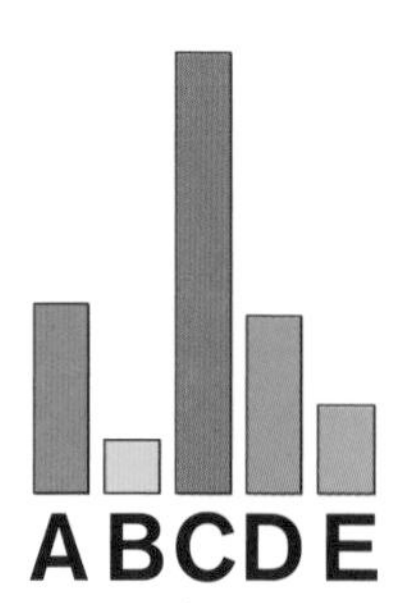

Reaction network

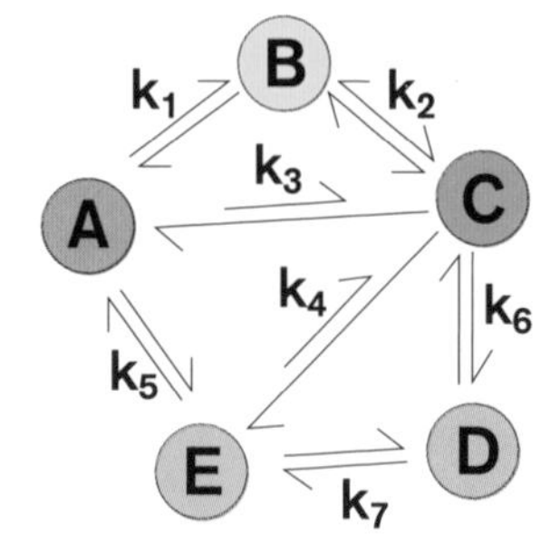

Energy landscape

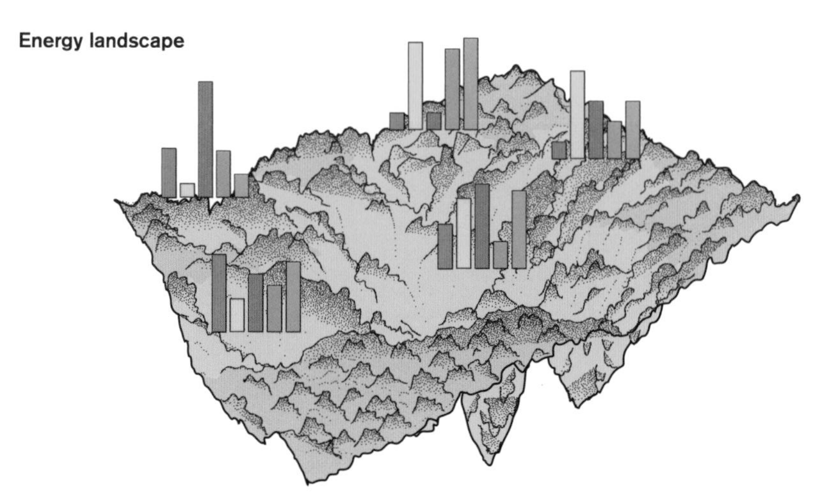

hyperdimensional maze, in real-time, according to externally imposed rules. At the end of the algorithm we could scan down the line of droplets and read off the chemical distribution in each droplet to obtain our final computed number.

The wedding seating problem

What kind of problems could we solve with a droplet chemputer? The wedding seating problem is a classic example. Imagine a wedding with 64 guests. We want eight tables each with eight guests, but we have a complex list of constraints about who can sit together! There are in practice up to 2,016 different relationships, but let's say most people don't care and we only have 102 pairs who absolutely must sit together, and 52 pairs who *absolutely cannot* sit next to each other. (Let's not speculate why!)

Embedding the problem

We can embed the problem into a mathematical representation known as a graph, where each person is represented as a point on a sheet of paper. We draw a green line between the points if they are friends and red if they aren't. Our objective is to split the nodes of the graph into eight groups of eight people, minimizing the number of red lines and maximizing the number of green lines within each group.

To solve the optimization problem, we assign *penalty* scores for every possible grouping of people so that bad arrangements get higher scores. We *add* a penalty point for each pair that *dislikes* each other on a given table. We *subtract* a penalty point for each pair that *likes*

Opposite: An imaginary droplet of water containing five different molecules (A to E), made up of atoms of different elements. The number (n_1, n_2...) of atoms of each element in the droplet never changes, but the number of each type of molecule in the droplet (bar chart) fluctuates over time as atoms move between molecules. The rate (k_1, k_2...) at which atoms move between molecules in the reaction network depends on the size of the energy barrier for each reaction. Many droplets like this could form a droplet computer with around 6×10^{23} atoms—approximately a teaspoon—of liquid. If we could read out how many of each molecule were in the solution at any one time—perhaps by spectroscopy (page 21) —that would tell us the "computational state" of each droplet. By tuning the energy landscape to a computational problem we could find the solution by reading the molecular breakdown of the droplets once they had reached a steady state.

each other on the same table. Finally, we build in penalties for groups that are too small, or too large. The score sheet we have described is a cost function, and it is analogous to the genetic and energy landscapes we explored in chapter 7. We can hunt through for seating plans with the lowest overall score. The score sheet nicely embeds our practical problem into a mathematical framework that we may encode in Von Neumann or chemical computers.

It is easy enough to solve this problem on a Von Neumann computer by scoring every possible combination, one by one. However, the time taken to solve this problem increases very quickly as the number of attendees, and so the number of possible seating plans, increases. For 64 guests, we would have to explore 4.5×10^{47} combinations.

The wedding seat problem is a familiar example that illustrates the kind of computations we might want to perform to explore molecular, genetic, or material landscapes. For example, if we used this approach to determine the 3D shape of a spider's silk protein, which has 3,000 amino acids that either repulse or attract each other, then we would have 4.5 million possible interacting pairs of amino acids, and computing the lowest-energy arrangements (equivalent to maximizing the green lines in our wedding problem graph), would take around a month. However, to really understand silk we would also need to model hundreds of copies of that protein interacting, which would take years.

Each droplet in our proposed chemputer is always trying to find its minimum energy arrangement. If we could set up the droplet array so that the energy landscape of the chemical reactions re-created the cost function of our wedding seating problem we could use this natural propensity of chemical systems to find the lowest score, and Aunty Gertrude won't have to listen to Uncle Bob complain about her cat all the way through the wedding. A computer architecture designed to solve advanced wedding seating problems more efficiently than a Von Neumann machine is precisely what our material-hunting AIs will need.

Opposite: A hypothetical droplet computer for a wedding seating problem with ten people, two tables, a number of good relationships (green), and a few bad ones (red). We need 20 logical variables (true or false) that represent whether each person is at a table in a given trial. Based on the interaction between guests, the number of people allowed at a table, and insisting that each person only has one seat, we compute a score sheet (analogous to an energy landscape). The information in the score sheet is encoded as colored couplings between droplets, while the state (1 = blue, 0 = orange) reports on the internal states of the chemical droplets, which then battle it out to find the minimum energy state that solves the seating problem.

Wedding guest problem

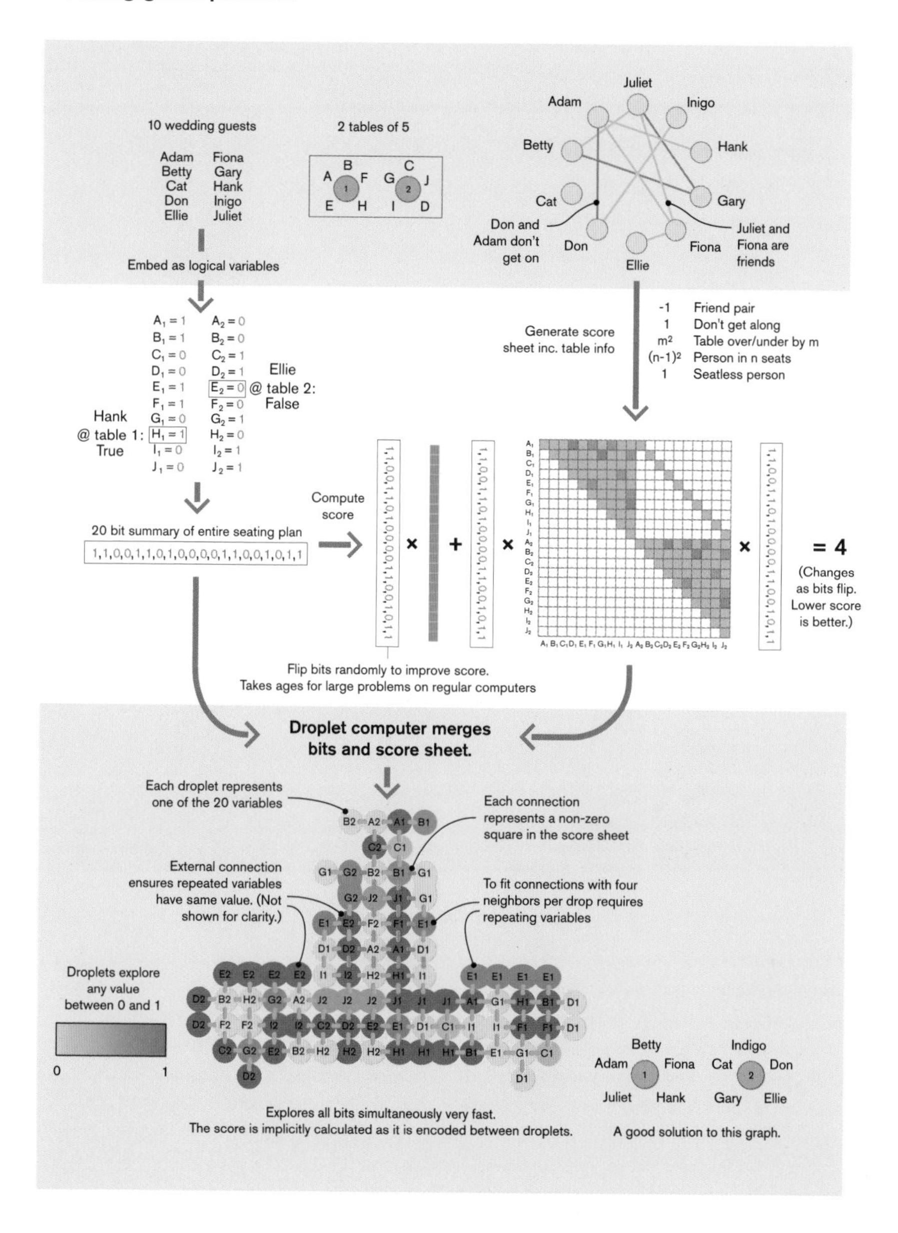

A real example of chemical encoding

One of the authors of this book (Forman) has been part of a collaboration between researchers from Wisconsin, Northwestern, Glasgow, and Toronto universities, to build a chemical computer that can solve this kind of problem.

Their approach is to set up chemical droplets on an array of electrochemical electrodes—each representing a variable in an optimization problem, such as the wedding seat problem. The number of hydrogen ions inside the droplet (the pH) can be set electrochemically and monitored visually with a colored dye. Electrical connections between droplets encode the penalty point scheme. When the system is energized, a chemical battle is passively mediated between the droplets. The cost function of our problem can be imposed by setting the strength of the connections between the droplets, which literally sets the energy barriers between the droplets. During the ensuing chemical battle the droplets change color because their pH changes, as electrons move around—not unlike the electrons of photosynthesis and respiration. At the end, the droplets take on the color that represent the optimal value of that variable in the solution. Because the variables are all calculated simultaneously—rather than in a series as with a conventional algorithm—the problem scales much better with the size of the number of variables.

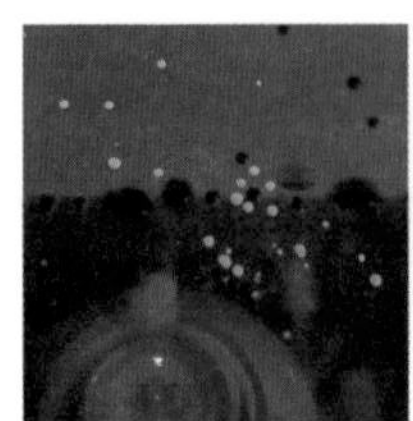
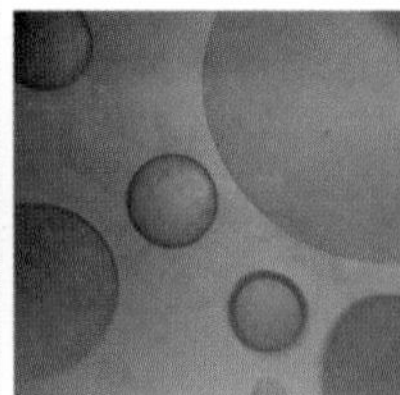
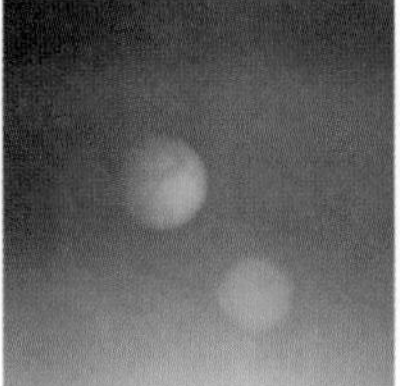
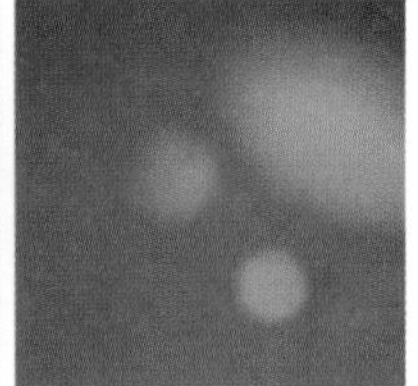

Above: Fluorescence can be used to reveal a droplet's state (left)—red for one, green for zero. Unlike the transistors in a microchip, our droplets can be both red and green at the same time, emulating some aspects of a quantum computer, but not all. Mixed droplet colors can be seen by viewing them through filters. With no filter (center left) we can see all the droplets; a green filter (center right) shows only droplets emitting green light; a red filter (right) reveals droplets emitting red light.

Exploring more complex geometries

Our chemputer is a non-equilibrium system (page 98) that dissipates energy continuously. Each droplet involved in the reaction must be kept in a steady state, otherwise the molecules will relax to their equilibrium state (page 98), so it is likely that they would all have to be energized at the same time. Some configurations of the reaction network would be able to dissipate input energy better than others. Therefore, varying input energy might change the probability of certain states emerging. This idea is called dissipative adaptation. Any changes to input must be synchronized between droplets, so state changes become correlated between distinct droplets, like a clock in a Von Neumann computer. By arranging adjacent droplets to exchange energy, altering each other's internal state—via light or chemical signals—the rules of the Turing machine for choosing the states of each droplet could be implemented between the droplets themselves. Each droplet effectively acts as the Turing read/write head for its neighbors in parallel.

By arranging droplets in a symmetrical lattice, we could create a scenario in which energy from one droplet could be transferred to multiple droplets far away down the notional Turing tape, creating complex networks of chemical droplets that talked to each other—a picture beguilingly like brain architecture itself, which requires complex graphs to map all the connections between neurons. Our computation is a pathway through the potential energy landscape of such a system to a state that corresponds to our computable number.

If the chemputer was a Turing mill then when the calculation has finished, we could remove the memory core of the chemputer and it would be our manufactured object! Maybe we could even make a regular computer processor this way?

Heterotic computing

A growing community of scientists is interested in how alternative forms of computation—be that quantum, biological, chemical or otherwise—relate to each other. They have even discussed the effects of combining two or more types of computer into a multi-physics processor, which is known as heterotic computing, a phrase coined by UK scientists Viv Kendon, Angelika Sebald, and Susan Stepney.

By building a distributed infrastructure made from chemical signals, we could link ourselves directly to the environmental world all around us. In the next chapter we will explore how such a breakthrough could change our future homes and cities.

Growing a control system

In the previous chapter we proposed that machine learning and simulations could help build our biosmartphone. Such algorithms could be run on conventional computers, but now we have a powerful new weapon in our thought experiment; a droplet computer! In our fabrication chamber we could deposit a droplet-based manufacturing computer that performed some or all of the simulations, experiments, and optimizations necessary to build a device. The design and fabrication process would merge, allowing the phone itself to take part in its own design, upgrade, and at the end of its life, its own disassembly.

In principle, a Turing mill *could* manufacture any computable object. If that is true, then perhaps the droplet computer could manufacture a regular computer processor directly into the mineralized synthetic bone infrastructure that supports our device. The first wave of our imagined biosmartphone fabrication process might be a Turing mill whose task was to construct a porous but solid structure similar to trabecular bone (chapter 4). A cortical bone outer surface of the skeleton could be deposited last to form a locally flat surface, embedded with small nanoparticles or molecules to generate electronic circuits. The Turing mill could use a generative testing algorithm that completes when the structure matches the specifications. Layers could be reabsorbed and deposited many times until it had computed a material arrangement that behaved like a Von Neumann processor. The final layer of the scaffold could be a layer of electrochemical interconnects between the CPU and the rest of the device.

Once the processing scaffold was finished and cleaned, a synthetic biology layer could envelope the construct in a tough cartilage-like matrix, trapping synthetic biological cells for the operational lifetime of the phone. The genetic circuits in the cells could continually re-generate and configure a chemputer next to the skeleton, as well as the outermost layers of the smartphone, just like our bodies continuously replace dead skin cells with new. This heterotic computing stack would combine a traditional Von Neumann processor for the day-to-day running of the phone with a chemputer for performing complex optimization algorithms and a synthetic biological layer for configuring the chemputer and interfacing with the external environment.

Below the transparent outer surface of the device, a "sub-dermal" layer might be able to manage the phone's appearance using an adaptive structural color photonic processor, similar to butterfly wings but also capable of performing optical computation and forming the final layer in the stack. This layer would consist of a mixture of light-sensitive cells for monitoring incoming light and reflective pixels that controlled the exterior appearance of the device. With such a spatially interleaved camera and display screen, you would be able to look someone directly in the eyes on a video call.

Beneath the screen a chemical energy system would power the lower computing layers by diffusing fuel through the phone, recharged by light hitting the device. The result would be a display screen capable of generating a range of colors with micron-sized pixels, whilst capturing energy and images. Pressure on the screen would create different patterns of light, adding responsive touch-screen functionality to our phone.

The biosmartphone

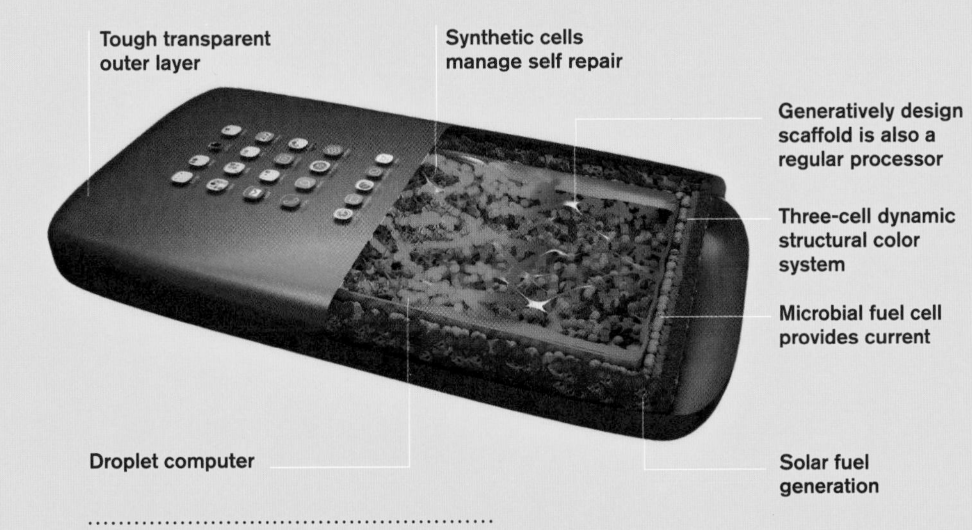

Above: Most components in this imagined device have analogs with constructs in biology. Putting them together with the new idea of a droplet computer could truly make something besides the sum of its parts.

9 The Cities of the Future

"Would the valleys were your streets, and the green paths your alleys, that you might seek one another through vineyards, and come with the fragrance of the earth in your garments."

The Prophet by Kahlil Gibran

Tree house

The technology that surrounds us in our homes is forever evolving. We have come a long way from simple farming homesteads, that supply inhabitants with basic needs, to buildings that boast connections to the global economy via multiple utility networks—sewers, water, electricity, gas, cell phone coverage, internet. With the addition of synthetic biology, additive manufacturing, and in-materio computation, it may be feasible to begin thinking about a new kind of domestic input system—a material supply network.

If we are to move closer to the concept of a circular economy, different sectors of industry must become more interconnected, with the surplus material from each process passing to the next. To facilitate these changes, a common platform for material sharing would help link industrial and domestic processes across the whole of human endeavor. Such a scheme might allow, for instance, car companies to trade waste more easily with, let's say, a clothing company, or an energy firm might find surprising sources of profit through commonalities with a water purification company.

Perhaps the technology for merging information and material that we have explored throughout the book could enable us to build an entirely new kind of network. One that connected residential and industrial buildings to a local reservoir of material resource and helped society manage its waste far more efficiently, as part of a global circular economy.

Network issues

At present, the concept of a material exchange reservoir to enable industrial "coincidence of waste" is little more than an idle fantasy. However, as we saw in chapter 3, some companies are beginning to implement similar ideas at a smaller scale—for example Wissington sugar factory (page 74) and Kalundborg Eco-industrial Park (page 76). At Wissington and Kalundborg both enterprises are engaged in what is essentially an efficient rearrangement of existing processes arising from a systems-based approach. The output products—food and low-tech applications—are not particularly sophisticated; rather, the novelty arises from the higher-order organization of the factories, which finds material efficiencies through co-operation and integration among multiple processes.

To achieve system-level thinking and link waste flows from different industries we need to take a step back to look at the main flows of material and energy around our society. We identify two major strategic problems caused by the architecture of our system: the last-mile problem, which arises from a *centralized* energy-generation and manufacturing architecture, and separation of supply networks, whereby major industrial sectors have separate but interdependent distribution networks, which increases the overhead costs of infrastructure.

The last-mile problem

Moving goods or services among a relatively small number of supply nodes in business-to-business transactions is straightforward. However, distribution over the last mile to millions of consumers presents a huge challenge—particularly for utilities, such as energy, water, internet, and mobile communications access. However, an emerging solution in the energy industry is a microgrid; a small, localized version of the national energy distribution grid, which provides power for individual buildings or a small group of buildings.

Instead of a huge remote power station at the end of a line of pylons, we have many small sources of energy distributed uniformly everywhere, such as solar panels, small in-stream hydroelectric turbines, or fuel-driven generators. At just the flick of a switch, such microgrids connect directly to the main national grid, which enables local users to sell energy surplus or buy energy from the national grid to make up for any shortfall they may have experienced.

Separation of supply networks

All the different supply networks that connect to your home and industrial sites are separate, but they operate using many of the same basic inputs, such as water and electricity. For example, water purification uses a huge amount of energy, yet vast quantities of water are needed to drive the turbines and cooling systems of power stations, and suitably treated wastewater would do just fine for cooling. In fact, many of the efficiency savings at Kalundborg stem from recycling of wastewater between industrial partners.

Similarly, agriculture requires huge amounts of energy to operate machinery and a large quantity of water for irrigation. Indeed, any industry with a distribution network will depend on some or all of the others. Internet, gas, mining and so forth. Pairing more of these

Above: Sewage and wastewater purification plants like this one in Denmark have to use large quantities of energy to remove impurities and produce clean drinking water, whilst power plants produce large amounts of waste water to generate electricity. Pairing these processes could yield significant improvements in efficiency.

industrial processes could yield huge benefits. For example, US company Cyclopure has developed a technique to filter out a wide range of impurities using cyclodextrin—rings of sugar molecules derived from corn. This process uses much less energy than reverse osmosis, which is the standard method used to remove water impurities and involves forcing water through a membrane against the direction it would naturally go. Moreover, Cyclopure's filtration system can be reused by rinsing it to remove filtered out materials, massively increasing the operational life of the system.

Merging solutions

What if the concept of the microgrid and Cyclopure's water purification solution could be combined? The capacity to *locally* purify a small reservoir of water would avoid having to transport it back to the water purification plant, thereby avoiding water loss in leaky pipelines and saving on energy.

We saw in chapter 5 the potential for synthetic organisms to metabolize and re-manufacture materials; this observation creates the opportunity to propose a *hypothetical* domestic ecology. A network of synthetic organisms, such as yeast and bacteria, could be engineered to metabolize organic molecules. Photosynthetic algae, such as *Volvox carteri* (page 132) could use solar energy to synthesize the sugars needed to fabricate cyclodextrin, allowing it to be grown in a vat, rather than a field. Furthermore, excess sugar from the water filter could be passed to a microbial fuel cell to produce electricity for the local microgrid. We could turn each of our homes into a mini-Wissington, in which algae digest the organic matter in sewage and wastewater, and manufacture cyclodextrin to filter and purify the water. Any organic waste from the algae could be used as fertilizer for gardens or as fuel for heat. Maybe the filtered solids could even be used as fertilizer to grow synthetic plants (or fungi) that could manufacture feedstocks for our 3D printer, adding yet more solar power into the mix. Any excess water could be stored in fungal cells, providing a reservoir for later use.

This integrated system would eliminate energy and water costs associated with growing corn, transporting and purifying water, transmitting electricity, and the industrial support industries needed to build pipelines and power stations. It would also capture CO_2 from the air and provide a local electricity supply. With the addition of a *single* common molecule at a strategic nexus, we could join previously distinct networks and achieve some remarkable efficiencies.

An integrated home

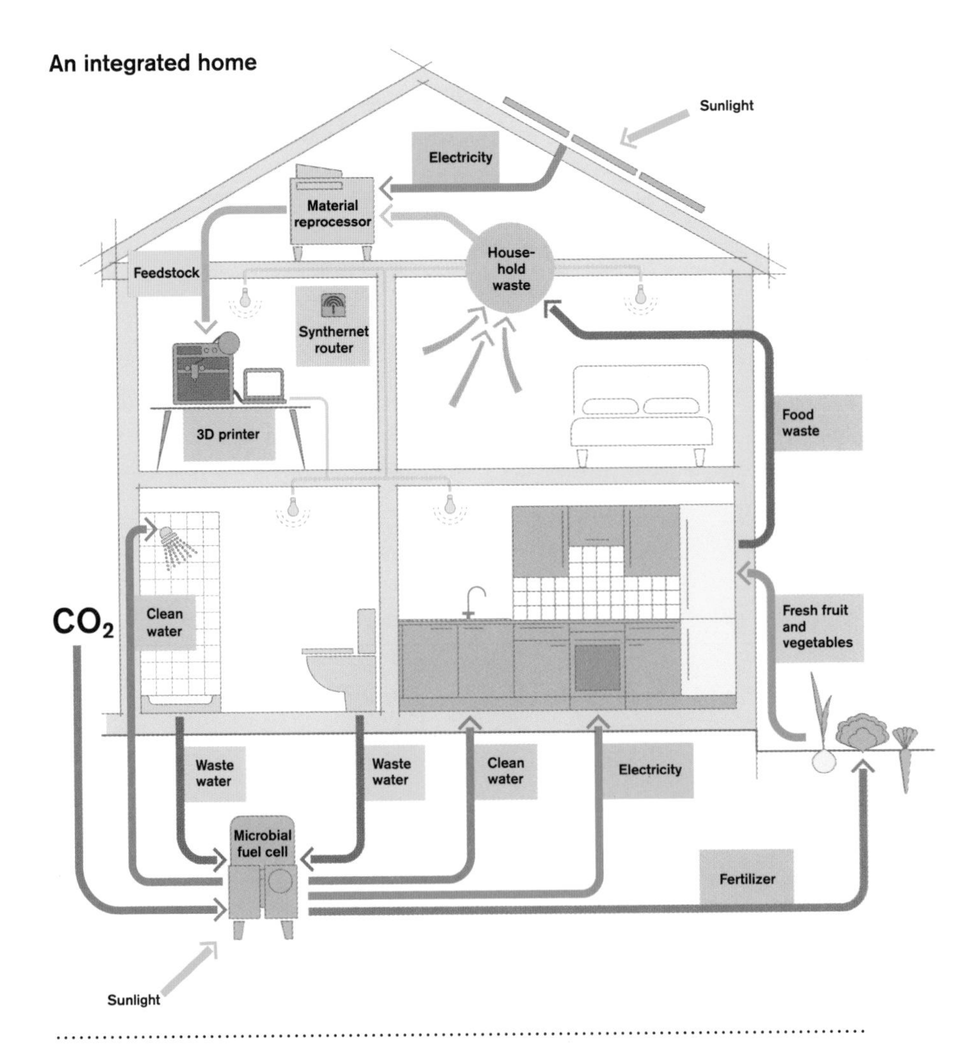

Above: Hypothetically speaking, a microbial fuel cell driven by sunlight and producing cyclodextrin would absorb CO_2 from the air, so our homes could become carbon negative. The microbes in these devices could potentially provide electricity, purified water, and fertilizer. If demand could be met this way then power stations and water purification plants could be eliminated, as well as the overhead associated with transporting electricity and water over large distances. Effectively, our house would play the role of a tree in the ecosystem.

Systems thinking

The idea of using a synthetic biology fuel cell to enable the convergence of a microgrid with water purification is purely hypothetical. However, it perfectly illustrates one of the key principles of circularity—the smaller the recycling loop the more efficient it is. In a linear economy, all the supply networks are separated and their processing centers—such as power stations and water purification plants—are spread far apart. The result is that we are locked into a system where wasted water is embedded into our energy generation and wasted energy is embedded into our water purification. Instead of using sunlight—a free local resource—to purify the water we already have in our pipes, we go ahead and build separate transport and processing infrastructures for the two commodities.

The technology we have outlined in this book has been strategically selected for its propensity to help us find emergent efficiencies, just like the cyclodextrin example. Nature is full of efficiencies that derive from the properties of individual molecules—the ribosome, DNA, ATP, and so on—all of which are used for a variety of applications. In the cyclodextrin example, the multiple possible uses of this single sugar molecule could enable the intersection of power and water flows, but only when the processing centers for those flows were co-located.

We believe the technologies we discuss throughout the book could help simplify the overall societal process of waste exchange between companies and find new routes for solar energy to enter the economy, besides crops and solar panels.

Landfills: the goldmines of the 22nd century

The library of technologies we have built could help merge existing resource networks. But we must add one new distribution network before we can make progress. In the current linear economy, we are accustomed to acquiring our materials from mines, but if that flow were to end, then where would companies go? As we have seen, some companies could partner up to provide some or all of their raw resources, but not every industry would be able to do this straight away.

A first port of call might be landfill. For example, there may be higher concentrations of gold stored within some landfills than in gold mines, but the reason we don't currently mine gold from landfill sites is that the mixture of materials is too complex to make it profitable. We have already invested time, energy, and cash into developing tools, some crude, some sophisticated, for extracting gold from the complex mixture of materials in ore, so it is always cheaper to keep relying on existing infrastructure to mine and refine raw materials from the ground. This is the reality for the majority of materials we use today, and of all the products we manufacture and use daily, only a tiny fraction—about 16 percent—of the world's waste is recycled. And as it becomes increasingly difficult to find sites where waste can be buried, many cities have resorted to burning waste instead of investing time, money, and energy into sorting and recycling. But in doing so they turn valuable raw materials into emissions, scattering resources even further afield as pollution—without even using the heat they produce usefully— adding to our problems.

As we saw in chapter 4, our current industrial mindset is to sort mixed material reservoirs into streams of pure raw materials, because our manufacturing is set up to receive pure inputs. At the macro level, robots controlled by trash-sorting AI could provide an economical solution to the purification problem. For example, the Sims recycling plant in New York operates Robb 2.0, a robot that can pick and sort 12,000 items from a conveyor belt every hour, and Apple Inc. has developed a series of recycling robots, the most recent of which is called Daisy. Daisy is a multi-armed robot that can disassemble nine different iPhone models and take apart up to 200 devices per hour, separating out individual components—around 0.67oz (19g) of aluminum, 0.028oz (8g) of copper, along with trace amounts of platinum, silver, tin, and gold per phone—making them easier to reuse or recycle than existing techniques such as shredding.

But what if we didn't have to regenerate pure streams of materials? What if we could manufacture directly from mixed material inputs like landfill, just as biological organisms do?

The biological solution

Yet again nature provides us with an example whose tremendous value is often overlooked: soil.

Just like landfill sites, soil is a complex mixture of many different elements, yet plants, bacteria, and fungi have evolved remarkably effective ways to extract the nutrients they need—with a little help from sunlight and air. Stable ecological communities generate as much soil as they use; otherwise they would exhaust their nutrient supply and perish.

Emulating the ability of biological organisms to manufacture with mixed material inputs would help businesses find material efficiencies and communities take the first steps toward local circularity. We could imagine a reservoir of material—generated locally from waste—that can be used as a feedstock for the synthetic biology driven additive manufacturing layer in a local stack of technology (chapter 5). This layer would

Soil: an exchange medium

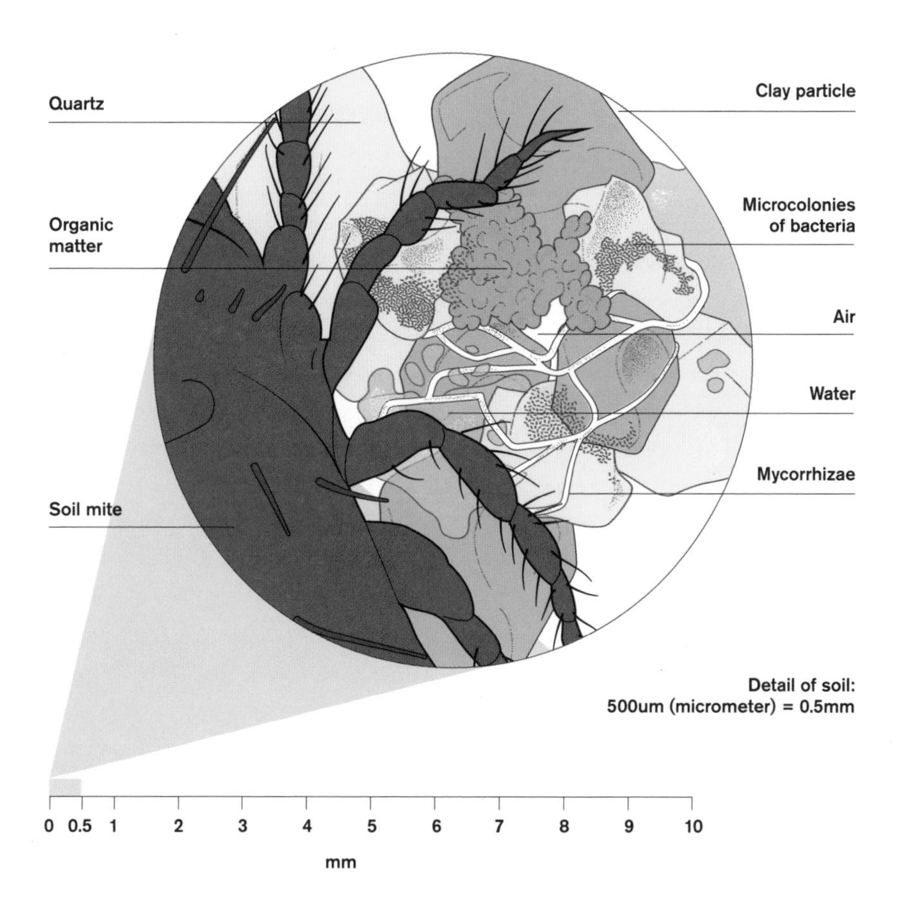

Above: Soil is an incredibly complex mixture of rock, clay, organic matter, and water that is teeming with biological organisms of many scales, from bacteria to earthworms. It is an exchange medium—an open market in which species from different biological kingdoms interact to trade waste materials. Nitrogen from the atmosphere is sequestered here and taken up by plants, before being passed on up the food chain.

automatically handle mixed inputs for processes higher up the stack, eliminating complex and costly sorting processes for waste material. Over time, the waste stream would have to become highly refined, but also remain dynamic, with optimal concentrations of different nutrients to supply medium-term changes in demand, just like the concentrations of sugar and oxygen in the bloodstream.

Synthetic enzymes that form self-regulating genetic circuits, perhaps integrated with in-materio artificial intelligence as part of a heterotic computer (chapter 8), could extract specific components from mixed feedstocks, such as landfills or industrial waste, based on real-time manufacturing demands. Perhaps our local infrastructure could reach down into a solid layer of such material—"technosoil"—underground, in a manner akin to the roots of trees, or it could be piped into our homes in the form of a liquid "city blood" via a pipe network that links all the houses in the neighborhood. The local bioreactor could use the nutrients to synthesize molecules the in-materio computer requested to complete its designated manufacturing task.

Indeed, in a remarkable twist of fate, researchers have found a way to use our favorite molecule—cyclodextrin—to extract gold dissolved in water. With a layer of technosoil beneath your home, you could just grind up your old phone with a home garbage masher and wash all the materials into the basement reactor for composting. As well as filtering your water, cyclodextrin could also filter out gold—and possibly other useful resources—from the technosoil mixture, to supply your house in-materio computer for the next device you ask it to make.

A network of in-materio computers

In-materio computers are perhaps the most powerful idea we encountered in all the technologies that we have explored. But just as the pioneers of the internet realized that the power of computers could be enhanced by connecting them, we anticipate that building a network of in-materio computers—which we call the "synthernet"—will increase their value substantially.

In previous chapters we have seen what is possible with the nanochemistry of biology. We have had a glimpse into the workings of ecosystems, in which species compete for resources but also co-operate and co-evolve with each other and their local environment. Companies that can figure out how to build in-materio computers, and subsequently how to connect them into networks, could create a material-centric internet that could form the basis of a merged microgrid of energy, water, information, and materials. Whereas the internet connects people to information, the synthernet connects people to the materials around them. Perhaps that is precisely the kind of interconnected material exchange platform that companies need to enable a circular economy?

Such a network could have extremely small recycling loops, leading to very efficient circularity. It could find and exploit emergent efficiencies that arise at the molecular scale when we begin to combine systems—such as combining energy harvesting and water purification in microbial fuel cells. A network of in-materio computers may integrate these kinds of processes into a single, albeit extremely complex, "system of systems" that would rival biology in its sophistication and form the backbone of a new kind of city formed from self-assembling infrastructure. Why invest over and over in renewing infrastructure—from roads and bridges to buildings—when you can invest once in infrastructure that repeatedly computes how to rebuild itself from local materials and energy that are already there?

Computation in every brick

Just as the proverbial jar full of rocks can accommodate additional sand, and the jar full of sand can accommodate additional water, the performance of vital tasks could move into the nooks and crevices of space around us. Imagine what we could achieve if the bricks used to build our houses and public spaces were crammed full of technology with biological levels of sophistication. Biological cells pack a great deal of functionality into an incredibly small place: as well as being the building block of a larger organism, a living cell can exploit energy stored in its chemicals, manufacture proteins, process waste, and construct new cells. Our homes might therefore take on more organic forms, taking inspiration from the branching architecture of corals or self-ventilating termite mounds. Every molecule of our buildings would have multiple functionalities, and with the concepts of in-materio computing we could perform all kinds of complex computations, even run sophisticated AI, through the very fabric of our homes and factories.

The synthernet

It would be tempting to think of the synthernet as just like the internet, and it shares some similarities, but it is fundamentally a different construct. The way that materials behave is completely distinct from the way that packets of information move around the web—we can't send molecules over wi-fi! Despite reaching the technological peaks of space flight, cloning, and self-driving cars, teleportation still eludes us.

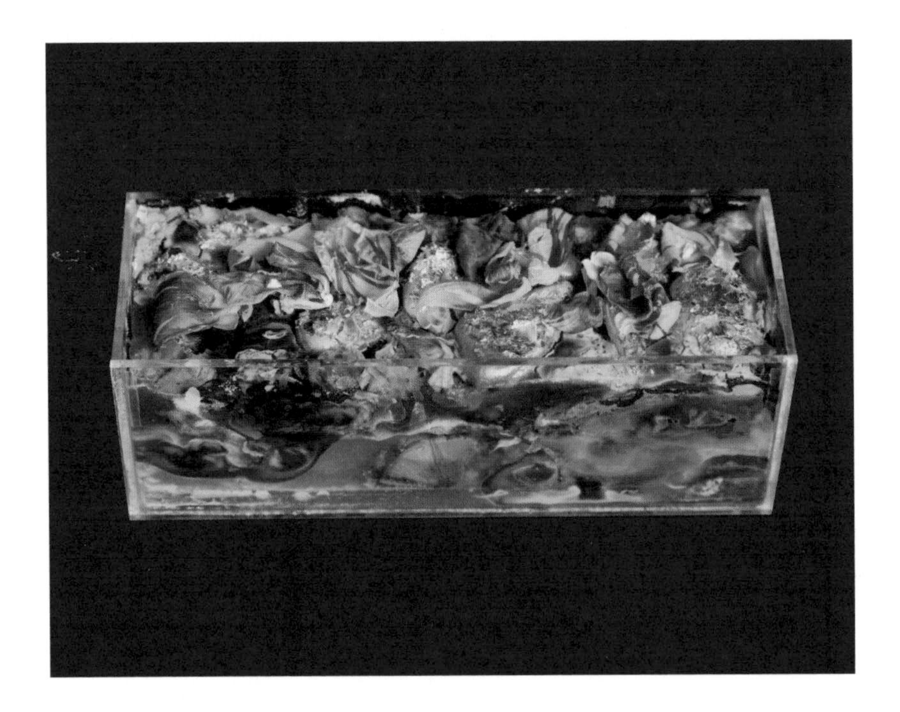

Above: A "living brick" designed by Professor Rachel Armstrong, an experimental architect at Newcastle University, UK, whose pioneering work has provided inspiration for this book. The brick-sized perspex box—created for the Adaptive Architecture Exhibition at Nottingham Trent University in 2018—exemplifies the idea of embedding chemical processes into the walls that surround us. The complex internal pattern is created by a cocktail of minerals, metals, and organic materials—including gold, pebbles, and tulip petals—that come together in a chemical "dialog."

Every web-browser on the internet uses the same system (HTML and TCP/IP) for accessing, downloading, and rendering data to present to the user. Such a standard guarantees that all users—be they domestic, commercial or otherwise—can interpret others' data, as long as their browser and computer obeys the standards. Such technology is built on the concept of a stack, such as the LAMP stack for web development (page 119), or the synthetic biology stack (page 123), and is highly reminiscent of trophic levels within an ecosystem (page 45). The synthernet would provide an interface between humans, the internet, ecosystems, and biogeochemical cycles.

Like the wood-wide-web (page 46), the synthernet would have a connection to a local reservoir of material where it could recruit or release resources, and it would need to communicate with neighboring chemputers to perform distributed analyses. If the internet is a tool that links humans to other humans, the synthernet is a tool to connect humans to their local environment and to the Earth as a whole.

A Synthetic Biology Additive Manufacturing stack could have the satisfying acronym SBAM. An initial design brief for the SBAM protocol might look something like this: The synthernet has two overall purposes; 1) to link humans to the matter they need to build their localized objects, be that matter drawn from a local material library, existing products, or the environment; and 2) make sure that the global sum of human emissions and chemical demands doesn't exceed the planetary boundary conditions (page 16).

To achieve these twin goals, the synthernet will need to act like a brake and an accelerator for the manufacturing process. The amount of energy available to the synthernet will be determined by local sunlight, just as it is for a tree. The sythernet will regulate the local rate of fabrication of food, devices, clothes, construction materials, and clean water, according to the amount of energy available, the resources at hand, and the priority of need (food and water before devices, for example). It would also need to manage recycling and waste exchange across the neighborhood.

In this scenario, each building would implement a stack of in-materio computing and synthetic biology protocols, enabling it to optimize resource acquisition and product fabrication locally—like solving the wedding seat problem for atoms in houses, instead of people at weddings. It would also be responsible for seeking inputs from experts and users on product design, resource limits, and other factors. Much of the detailed design of objects could be carried out by conventional AI.

Every home should have one

An early form of such technology might take the form of material-reprocessing units that can be installed in a basement, or built into the walls of new homes, as bolt-ons to the existing infrastructure. These material reprocessors might be quite simple—processing biological material much like a home-brew kit, or the coupled microbial water purifier and fuel cell discussed in the opening of this chapter—but could supply much of the resources for our activities.

The entire system would be powered by sunlight, while synthetic biological cells could perform the necessary chemical reactions to produce food or build items. Our diets could be a rich mixture of vat-grown printed products—similar to many of today's meat substitutes—fresh fruit and vegetables grown on vertical hydroponic farms encasing our homes and restaurants, and—if you like—local free-range meat and dairy products harvested sustainably from our surrounding natural ecosystems.

Waste materials or surplus production could be reprocessed on site or transported over short distances and sold to the nearby community. Information, transmitted via regular internet, might be translated into DNA for storage in the home chemputer, or into RNA to produce proteins that will execute the script.

Storing and recycling water, energy, and material resources locally would mean that instead of garbage trucks collecting our waste we could recycle *everything* locally, which is going to save a heap of cash! As home recycling systems follow typical technology improvement curves, systems such as the one we've described might one day cost just thousands, or tens of thousands of dollars. If it meant you could recycle 80% of your materials and buy 80% less from stores, then it might be worth investing in. Why buy new atoms if you can just reuse the old ones? Of course, not everyone has the upfront capital to make these kinds of investments, so government loan schemes could improve uptake of new circular economy technology considerably.

Supporting infrastructure

With our farms and factories integrated into the building blocks of our homes, we would be free to design our above-ground infrastructure around the needs and preferences of people, and design synthernet infrastructure to efficiently supply city blood, or tap into technosoil, throughout the city.

The internet protocol: a classic model

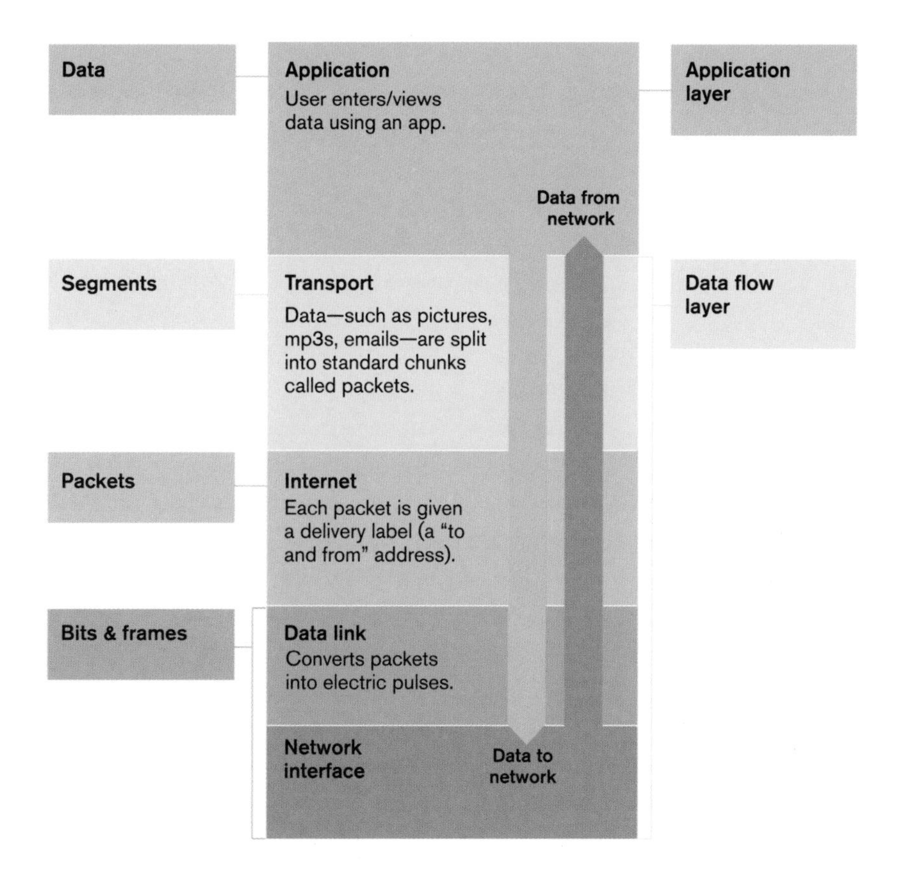

Above: The Transmission Control Protocol/Internet Protocol (TCP/IP) reference model is a representation of different layers of functionality that computers connected to the internet use to communicate. The user interacts with the top layer and each subsequent layer performs a vital task that ensures data transfer is reliable. The transport and internet layers handle addressing and delivery; the link layer converts digital data into physical signals, such as light or electric pulses, for transmission over the network.

With self-sufficient local supply chains, our need for transport would be almost entirely restricted to moving humans around—to school, work or places of recreation. Communications technology could also be embedded into the very fabric of our homes, aiding the current trend toward home working, reducing stress on our transport networks, and allowing for greater flexibility in local and national infrastructure designs. Given these new possibilities, we could again take inspiration from nature to optimize our cities of the future for efficiency and aesthetics. As we have seen with the examples of ants and neurons, living organisms are excellent at solving optimization problems, and researchers are exploring the use of simple microbes such as slime molds to improve transportation networks. Such strategies could help us plan the above- and below-ground infrastructure of each synthernet city, from roads to underground pipe networks.

The synthernet and economics

A key principle of the circular economy is that it operates in conjunction with a free market—where the price of different commodities is determined by supply and demand. In the same way that microgrids can import or export energy from the national grid, users could import and export local waste at the current market rate. Indeed, if such localized recycling was taken as far as it could go, then eventually it might even be possible to start listing different categories of waste as commodities on a material stock exchange—industrial systems would need different mixtures than domestic. Eventually, once we have become fully circular and the majority of virgin raw materials have been replaced in favor of mixed recycled materials, perhaps the stock market would only deal in different categories of waste, rather than individual purified materials such as copper and zinc.

With the right technology, by saving up all the trash that goes through your house and stock piling atoms locally, that garbage suddenly becomes an asset, rather than a burden. It becomes capital. In fact, a benefit of the circular economy is the build-up of different forms of capital—economic, natural, human, social and financial capital. If you capture and recycle materials locally you can build assets while supplying yourself with energy, clean water, garden fertiliser, and building materials, then with even the smallest amount of natural capital, you could cover most of your basic (and more frivolous) needs.

You could then trade any surplus material and energy derived from that capital across regional and national networks, providing some revenue that you could use to buy anything not supplied by the local material reprocessor. Individuals could channel surplus energy and

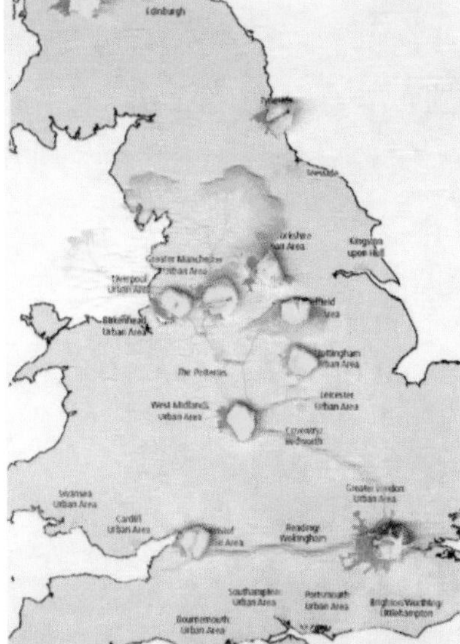

Above: Slime molds like *Physarum polycephalum* form a fungus that grows on rotting vegetation, which searches for food by extending long filaments out to randomly search the space, forming permanent structures via positive reinforcement. Researchers at the University of Bristol grew *P. polycephalum plasmodia* on maps of the USA (top) and the UK (bottom), with oat flakes at the sites of major cities to visualize efficient transport networks. They found revealing similarities—and differences—with reality.

materials to larger synthernet centers capable of building vehicles such as buses or self-driving cars. Suddenly, we are all trading local imbalances in particular elements and energy across the supply networks, and cash is no longer the primary goal, unless you want it to be.

Such a decentralized model of manufacturing would only work if the marginal cost of channeling the flows of matter and energy into a new product was less than the cost of making that product in a dedicated factory and then transporting it over a long distance. An economic break-even that is practically guaranteed, given that the cost of overheads per product shrinks to zero as the number of recycling loops increases, with the transactional costs paid for by the conversion of sunlight into earthshine. Ultimately, products of such a system might even become free at the point of use—in the same way that a monkey in the jungle doesn't pay the tree for fruit—at least not with cash.

Inevitably, our relationship with materials, and with our homes, would change. We would need to find ways to make use of surplus supplies—we could no more turn off our house than we could switch off a tree or the Sun, so our homes would be continuously producing resources whether we used them or not—the synthernet is an algorithm that never completes (page 164). Just as jungles grow constantly, our new urban jungle would yield endless supplies of materials that might exceed our current levels of consumption. Such surplus could even be used to build the neighbor's reprocessor for free, resulting in the rapid spread of the technology.

A question of scale

What it would take to build a city on the principles outlined here?

There are some simple physical scaling laws that can help us explore what the limits on cities and planets might be, and what this means for the quantity of matter that could be processed in a single city. From this, we could calculate the maximum amount of energy and materials that each person could consume on average and note that it compares favorably with what our current systems provide (page 223).

In his book, *Scale*, theoretical physicist Geoffrey West points to an innate advantage in living in groups: the living cell in an organism gains more return for the same expenditure of energy. This fundamental principle has helped shape the evolution of multicellular organisms and the colonies of superorganisms seen in ants, wasps, and bees. However, this principle is bound by the law of diminishing returns. Ten cells cooperating may achieve a great relative advantage compared to one cell, but adding the

See the wood for the trees

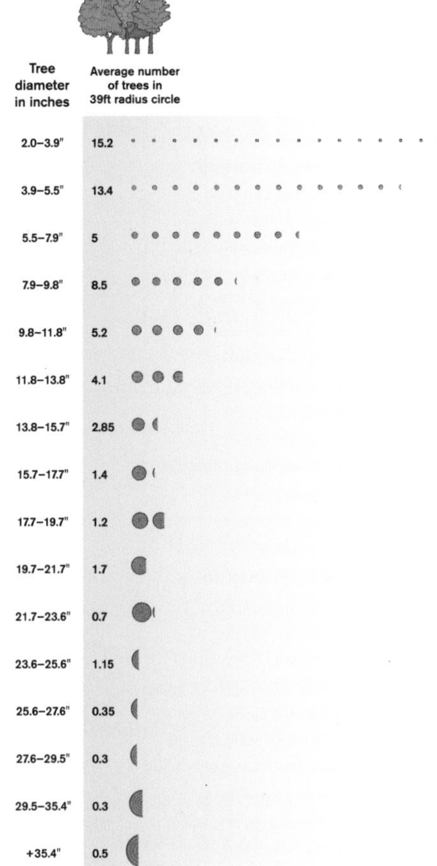

Above: The variation in size of different trees in the forest tends to follow a predictable pattern—many small, lots of intermediate, and just a few very large trees—and each type of tree occupies a slightly different niche within the woodland ecosystem. Similarly, different industries within the synthernet circular economy would need to operate in different niches and at different scales. Indeed, as people bought or sold matter via the synthernet their capital store would grow or shrink and eventually we would have a distribution of different sizes of capital much like the one shown above.

ten trillionth cell probably doesn't make much difference to the collective—although it would for the cell. Similar arguments tend to apply to city-dwellers: it's easier for a human to survive in a city where food and services are readily available for comparatively little effort, which frees time for work and other endeavors, resulting in a boost to productivity. Presumably a scale limit exists for technology as much as it does for bees and humans in cities, which if formalized could provide an estimate of the density of technology that is possible in a building, and the amount of energy available to run such technology.

The early stages of this transition are already apparent. Rather than creating universal access to information, synthetic biology companies are beginning to create universal access to materials and using biotechnology to create commercial access to the same kind of nanotechnologies found in the natural world. As these networks of companies begin to find new products and synergies in materials and waste exchange processes, they will leave others behind.

Entering a new epoch

We are social animals and the way in which we live together and the rules that we set for ourselves cannot be isolated from our technology, our cities, our countryside, and our wild lands. We are beginning to learn that there are global rules that we have no choice but to obey, for such rules are determined by physical and biological realities. We cannot change the laws of thermodynamics, or alter the biological laws of scaling, but nature shows us that we can adapt to live within them. The synthernet promises to revolutionize our manufacturing, industry, and commerce, and in doing so resolve many of the 21st-century conundrums that face our species, from climate change to environmental pollution, food security, and inequality.

With this overarching concept in mind, in the final chapter we consider the broader implications of a synthernet-based circular economy—how it will influence our societies and economies, how it will change our relationship with nature, with each other, even with ourselves. And we will see how the technologies we have explored could ultimately pave the way to human civilizations anywhere in the solar system, or beyond.

Our materials allowance

Returning to our back-of-the-envelope calculation from chapter 1, recall that just 0.05 percent of the Earth's surface would need to be covered in photovoltaics, or synthetic photosynthesizing organisms, to produce the 27 TW of energy humanity is predicted to need by 2050.

The total land area on Earth is 57.3 million miles^2 (148.3 million km^2)—about 29 percent of the surface area—the rest is covered in water. If we assume that the human population will peak in 2050 at about 10 billion, then the area per person is 159,600ft^2 (14833m^2) (400ft x 400ft/122m x 122m). Consider that an average home might be around 1100 to 3200ft^2 (100 to 300m^2). This means we could donate 0.15 percent of our land allowance to generate the 27 TW we need, 20 percent for infrastructure and community industries, and 60 percent for natural ecosystems, and still allow ample space—approximately 32,000ft^2 (3000m^2) per person—for our homes, gardens, and recreational spaces.

How much matter would be available to our home bioreactors? In 2017, humans extracted an estimated 101 billion tons (92Gt) of raw materials from the Earth and recycled about 9.53 billion tons (8.65Gt). Add this to the 1 trillion tons (900Gt) estimated to be currently tied up in long-term use, such as buildings and roads, and we have a total global stock of roughly 1.1 trillion tons (1000Gt) of material. Around 62 billion tons (56Gt) pass through individual hands each year in the form of short-term products that are either wasted, incinerated or, occasionally, recycled, giving each person on Earth a current consumption rate of roughly 8 tons (7t) per year (although in reality this consumption is heavily skewed toward people living in developed nations).

In our vision of the synthernet, all matter would be recycled within global material flows—some lasting months, others decades—meaning that we might have different material allowances for different types of matter. If we stopped mining new resources tomorrow, the materials we have already extracted could provide 10 billion humans with approximately 105 tons (96t) of matter each for everything from their day-to-day activities to the fabric of their homes, roads, and industrial hubs. If we consider that this matter would be combined intelligently to produce materials with emergent properties that can respond to the environment and even perform calculations, then this allowance—although lower than our current consumption—seems very generous indeed.

Ordering

In these case studies, we've looked at how a biosmartphone might look and function, but how would its fabrication and ordering fit in within our cities or rural homes? Ordering via the synthernet might—superficially—be like online shopping today, but the manufacturing process behind the scenes would be very different. A user could log on and browse the available smartphone designs before selecting their favorite and downloading it. From there, they could tweak the specifics of the phone, giving much greater control over the products they create. Once chosen, your design would be submitted to the synthernet ordering system for production.

At this stage, the process becomes radically different from what we know today. With a heterotic computing and manufacturing stack all around us, which we have called the synthernet, your home and neighborhood would literally be your farm and factory. After placing your order, the layers in the synthernet technology stack would kick into action.

In the context of the synthernet, a product is not just a product, it's a vital part of the economic ecosystem and occupies a trophic level. When you order a product, the fabrication process must fit that product within the overall economic energy flow and balance of materials in the synthernet ecosystem, so there would be limits to how much of a given material would be available in the neighborhood each day. Different manufacturing tasks—food, your new biosmartphone, your neighbor's new dishwasher, or stationery for your local school—would need to be prioritized and queued to fit within the available supply of materials and energy. The entire infrastructure around us would be continuously computing algorithms to optimize supply to meet the ever-changing demands of the users—the users in this case being natural ecosystems and even artificial consciousnesses, as well as humans!

The synthernet algorithm could detect the parts on the verge of failing and request a manufacturing slot from the house or neighborhood material reservoirs, without you even knowing about it. Like an automatic software update, our phones of the future may send us periodic notifications that new parts are ready for installation. Worn-out components could then be automatically composted in your home material re-processor, where the basic building blocks could be used to create new items, or sold into the network for other users to request, completing the circle of the synthernet economy.

The result would be a smooth technology stack of computation allowing a seamless connection between the human that used the handset and the biochemical, biological, and technological worlds around them. We would have a continuous process of never-ending production that was largely automated that we would simply direct from time to time, like tending a garden. The synthernet would also be able to learn and adapt to our anatomy and physiology and produce bespoke medical diagnostic devices that would provide us with continuous personalized medical care. Over the years it is possible that humans, AIs, and their self-optimizing, ever-learning, chemputing, manufacturing infrastructure would become indistinguishable from each other and the local natural ecosystems in which they were immersed.

Above & right: Full Grown Ltd is an art project that redefines our commercial relationship with nature. It takes only six years and patience to grow willow trees directly into chairs. The solar powered process needs no electricity, factories or mills, emits oxygen, absorbs carbon dioxide, and creates habitat for birds, bees and other wildlife. Each unique chair could last for generations or be recycled carbon neutrally at any time.

10 Taking it Further

"Take dramatic technological change as an invitation to reflect about who we are and how we see the world."

Klaus Schwab, *The Fourth Industrial Revolution*, 2017

Orbital habitat

On this epic scientific journey to find our *Brave Green World*, we have marveled at the torrents of sunlight cascading into our environment, ripping electrons from molecules, and triggering successions of chemical reactions. As terajoules of energy flood into the world every second, they senselessly batter molecules, vibrating chemical bonds, and ultimately return to space. Zooming out to the macroscale of entire organisms, we have observed the destructive capabilities of insects' relentless mandibles and wondered at their ability to construct vast self-sufficient underground civilizations. We have witnessed the importance of the most diverse ecosystem on Earth—the tropical rainforest—home to millions of species, all in constant competition for space and resources, and yet all interdependent in the cycle of life. We have seen how our own species, by tapping into high-intensity energy sources such as coal and oil, has been able to modify materials and environments at speeds and scales unmatched by any natural ecosystem—and how this ability has lulled us into a false sense of superiority and infinite growth. We looked to the natural world to understand how to control the flows of energy more efficiently and ways to manufacture more ingeniously, while maintaining the quality of the essentials for our existence—soil, water, and air—and replenishing the global reservoirs of material we depend on, so they remain forever intact.

Global localism

Perhaps in the future we can build our buildings from the ground up on a ubiquitous stack of technology inspired by living organisms, which connects the power of computation with chemistry, forming the foundations of a global circular economy. The synthernet would manage the local connection to the supply networks that sum to form our global economy, and in doing so it would create a two-way biochemical connection between individuals and their local ecosystems. Collectively, this could add up to form a well-managed bi-directional interface between the global economy and the flows of material in the natural marine, atmospheric, geological, and biological worlds, allowing humans and the planet to exist in symbiosis.

It is now time to take stock and reflect on these ideas, and understand how they might affect us, and whether this is a vision we wish to pursue. However, our vision is not static. These are not instructions carved in stone, but a framework that can be adapted, molded, and evolved until it functions. The concepts and ideas outlined in this book can be applied in many ways. It is up to us all as individuals to decide how our own ideas fit into the bigger picture and communicate our needs and opinions to others. Simultaneously, we must empower ourselves to listen to what others have to say. Only then can we construct a shared vision for humanity that works.

The individual in the collective

At its heart, the synthernet is a tool for managing that age-old battle between the individual and the collective; the idea that we can act as individuals for our own benefit and completely ignore the consequences for our economy, societies, and ecosystems that together form the complex surface systems of our planet. Like it or not, we are all simultaneously of the individual and of the collective. The ant colony that finds food isn't lucky—it almost guaranteed its success by sending forth a party of ants that were competent to explore the entire space and take advantage of whatever they found. Moreover, there was a simple mechanism in place to help the rest of the colony benefit from the good fortunes of each ant. The ants who didn't find anything weren't lazy; they made the lucky ant's job easier by narrowing the area the lucky ant had to search. The success of the colony emerges from a deep understanding of how individuals connect to the group.

So it is with humanity and the global economy. Every person is like an ant exploring facets of human existence, and the subconscious rules that we each follow give rise to the emergent properties of the collective. But it is not always clear which rules and policies give the emergent properties we are looking for. When there were few of us and we were all spread out, it hardly mattered to other groups what each collective looked like. Now there are many collectives and their boundaries are blurred. All our groups have merged to form a singular collective of collectives and the resulting system is extremely complex.

Different people have different ideas about what the global and national collectives should look like. The idea of the synthernet is a flexible tool that would enable us to capture the demands of the individual and groups, and explore useful emergent phenomena at the same time managing the sum effects of those demands on the global systems of economy and ecology. Implementing the synthernet would offer us a chance to manage that complexity and adjust the sensitivity of the global feedback loops to individual behavior, giving us a tool to maximize individual choices while minimizing collective impact. This will enable us to bring current overshoots back into balance, imperceptibly and gently over time.

It would be arrogant to think that we could take complete control of natural ecosystems or geological processes—this view has led to countless problems, from forest fires to invasive species—but the synthernet ensures that our own technological and synthetic biological ecosystems can fit into the larger global system harmoniously. We need to learn good rules, which we can only do by accepting and understanding that what each individual does has a collective impact; once we have accepted that, the synthernet becomes a sandbox that allows us to explore that world, carefully.

Nomadic ecosystems

As genomics and synthetic biology have drilled down into the minutiae of the workings of biological systems, ecological modeling has zoomed out to give us a broader understanding of how ecosystems function and how energy and nutrients flow between species. By learning to construct entire industrial ecologies, and teaching their operation to our descendants, we will discover and pass on the necessary tools and understanding of how natural ecosystems function, and perhaps we will reconsider the boundaries between our own and the natural world. In that case we may reach a new technological epiphany. Perhaps then we could become nomads not farmers once more. Instead of moving between ecosystems and adapting our lives to suit them, like our hunter-gathering forebears, perhaps we could take our ecosystem with us. Nomadic ecosystems!

We can think of no better way of describing this than to imagine a hypothetical excerpt from a 22nd-century textbook.

Excerpt from a 22nd-century textbook

Toward the middle of the 21st-century, synthetic biology, nanotechnology, and additive manufacturing had converged on a powerful set of automated technologies that transformed human society as much as farming had done ten thousand years before. With the breakthrough afforded by ubiquitous heterotic chemputers, domestication of these powerful technologies began in earnest around 2050, and just as home computing in the 1980s gave laypeople access to professional tools, large parts of the manufacturing process moved from remote factories and farms directly into our houses. Suddenly, everyone could make and recycle much of what they needed locally, from clothes to food to computers.

Our 22nd-century ability to design ecosystems wherever we go allows us to draw sufficient surface matter, for a given population size, from almost any environment, and to reprocess that matter indefinitely. Wherever we go, once we have taken what we need to secure our existence, we need take no more. Our populations can hunt and gather whatever they need from within a single settlement, starship, or city with balanced matter-exchange between the habitat and the local natural environment. Ultimately our new abilities have broken our dependence on land and farming, and in doing so, shattered one of the hallowed pillars of second-millennia economics on which the puny wealth of nations was built.

The nanotechnology on which the wealth of planets is constructed has a considerably larger surface area than the planets have real estate. A realization, which, almost incidentally, powered our journey to the stars.

Managing biological and technological flows

As we traverse the solar system we are learning more and more about geochemical cycles (page 70) and discovering that these cycles are as important on other planets as they are on Earth. Pluto, for example, has a pulsating cycle of nitrogen as the planet

Above, top: An algal bloom—caused by excess nutrients such as nitrogen in the ocean—swirls around the Swedish island of Gotland in the ocean currents of the Baltic sea, revealing the chaotic and beautiful patterns it creates.

Above: As sunrise hits the Sputnik Planitia basin, temperatures rise and the surface of Pluto's nitrogen ocean vaporizes into gas. About six and half Earth days later, as the sun sets on the Plutonian day, the vaporized nitrogen condenses back into ice. This artificially colored radar image taken by NASA New Horizons shows variations in Pluto's surface characteristics—evidence that once, rivers of liquid nitrogen may have flowed across the surface.

orbits the sun. This continuous cycle is central to Pluto's weather patterns, powering the great winds that blow across Sputnik Planitia, the frozen ocean of nitrogen on Pluto's west side, and shaping unusual plains and dunes that stand out against the rest of Pluto's surface. It also creates the unusual feature of high-altitude winds that blow in the opposite direction to the planet's rotation on its axis.

In chapter 8 one of the ideas for a chemputer was based on driving oscillating reactions in droplets, which are essentially miniature versions of planetary material cycles. The hallmarks of such energized feedback loops, where the output becomes the input, is that the output can be disproportionate to the input, amplifying the signal. These kinds of periodic processes, like the weather, have complex repeating and chaotic patterns that are often computationally unpredictable over a long period of time, and sometimes they can oscillate out of control—the same effect that occurs if a guitarist holds their electric guitar too close to the amplifier, resulting in a squeal of feedback.

In a nomadic ecosystem it will be up to us to decide the level at which each element or compound will circulate—known as the set point—as it moves through the biological and technical flows of the synthernet. When the first nomadic ecosystems are built, we may have to correct some overshoots that we have inadvertently created through our linear economy over the last two centuries. For example, we might command our local synthernet to imperceptibly reduce the amount of carbon circulating in the atmospheric stage of its global cycle by storing it in rock, plants, water, or even the fabric of our buildings.

The most low-tech solution is storing carbon in biological matter—increasing the global forest cover by 25 percent, for example, would capture an estimated 220 billion tons (200Gt). CO_2 can also be liquidized and injected into porous rocks such as basalt, where it dissolves in groundwater and reacts with the rock to form carbonate minerals.

But why bury such a useful element? Carbon has become a dirty word in current society, but diamond and graphite demonstrate the diversity of materials that can be produced from this versatile element. Direct Air Capture (DAC) technology uses a continuous flow of air through potassium hydroxide-filled towers to capture carbon dioxide, which can be converted into fuel. We might even print with carbon. CO_2 can be converted into portlandite an oxide mineral—dubbed Co2ncrete—that offers the construction industry an eco-friendly substitute for concrete.

Once we have brought our carbon, nitrogen, methane, and other global elemental cycles within the desired ranges, we will need to closely monitor every aspect of our nomadic ecosystems to ensure they remain in balance—just as a thermostat constantly regulates the temperature in your home to keep it within the set range. AI and chemical computing will give us the tools we need to monitor the positive and negative feedback loops in our technological flows and how they interact with the ecosystems around them, giving us precise control over our planetary material flows.

With industrial and domestic activities shaped to fit harmoniously alongside biological material flows within the resource limits of our planet, we will be able to locate ourselves anywhere there is energy and material to manipulate. At this point, the route to the stars is well and truly open.

A pathway to the stars

No matter which direction we take in the future, we face the same challenge—how do you build a closed-loop recycling system? Whether we stay on Earth, build habitats on other bodies in the solar system, or construct vast interstellar generation ships that must carry everything with them for centuries of travel to other stars, we must solve the same recycling problem. With few exceptions it would be foolish to try to carry all the necessary materials and tools to assemble and maintain colonies on other planets, not least because our solar system teems with all the building blocks we might need to build our own ecosystem, such as oxygen, nitrogen, potassium, calcium, aluminum, nickel, cobalt, and titanium. So, what if we could simply take a 3D printer and a resource-mining machine and build our homes and factories from the local resources?

Scientists are already experimenting with the challenges of fabricating bricks using Martian soil, or 3D printing with ice to build short-term structures in which humans can live and work. A few self-assembling structures could provide a base from which robots can source materials and build more permanent living and working quarters before any human ever sets foot on another planet.

The immense challenge of space travel has driven us to create our most sophisticated design and engineering innovations, many of which have proved their utility on Earth, too. Polyimide foam for lighter insulation, methane detectors, miniaturized computer chips, emergency medical "space blankets," and infrared thermometers, all derive from technology developed for our exploits in space. Software written for lunar image analysis helps interpret data from medical diagnostic tools such as computer-aided tomography (CAT) and magnetic resonance imaging (MRI), and "cool suits," developed for the

Apollo moon mission have been adapted for use by racing drivers and nuclear reactor technicians. There is a beautiful symmetry, therefore, in the idea that the limitations we face on Earth might lead us to design technology that could pave the way for civilizations elsewhere in the galaxy.

What future do we want to live in?

The *Brave Green World* we have hypothesized is not intended to be a utopia, but a realistic and sustainable future toward which we can all strive. A circular economy would solve many of the issues that we are facing today, but it would also generate new ones, which we must be prepared to accept, mitigate, or plan for. In developing nomadic ecosystems, we will fundamentally alter our relationships with the natural world, with matter, money, information; it may even alter our own identities. It is beyond the scope of this book to consider all of these issues in great detail, but we will consider some of them and the adjustments that might come hand-in-hand with a future founded on nomadic ecosystems.

Above: Direct Air Capture (DAC) technology allows air to flow through potassium hydroxide-filled towers to capture CO_2 from the atmosphere. At this garbage incinerator near Zurich, Switzerland, captured CO_2 is passed on to a nearby mineral water manufacturer for manufacturing carbonated drinks, and fed into greenhouses to increase crop yield.

Synthernet global flows

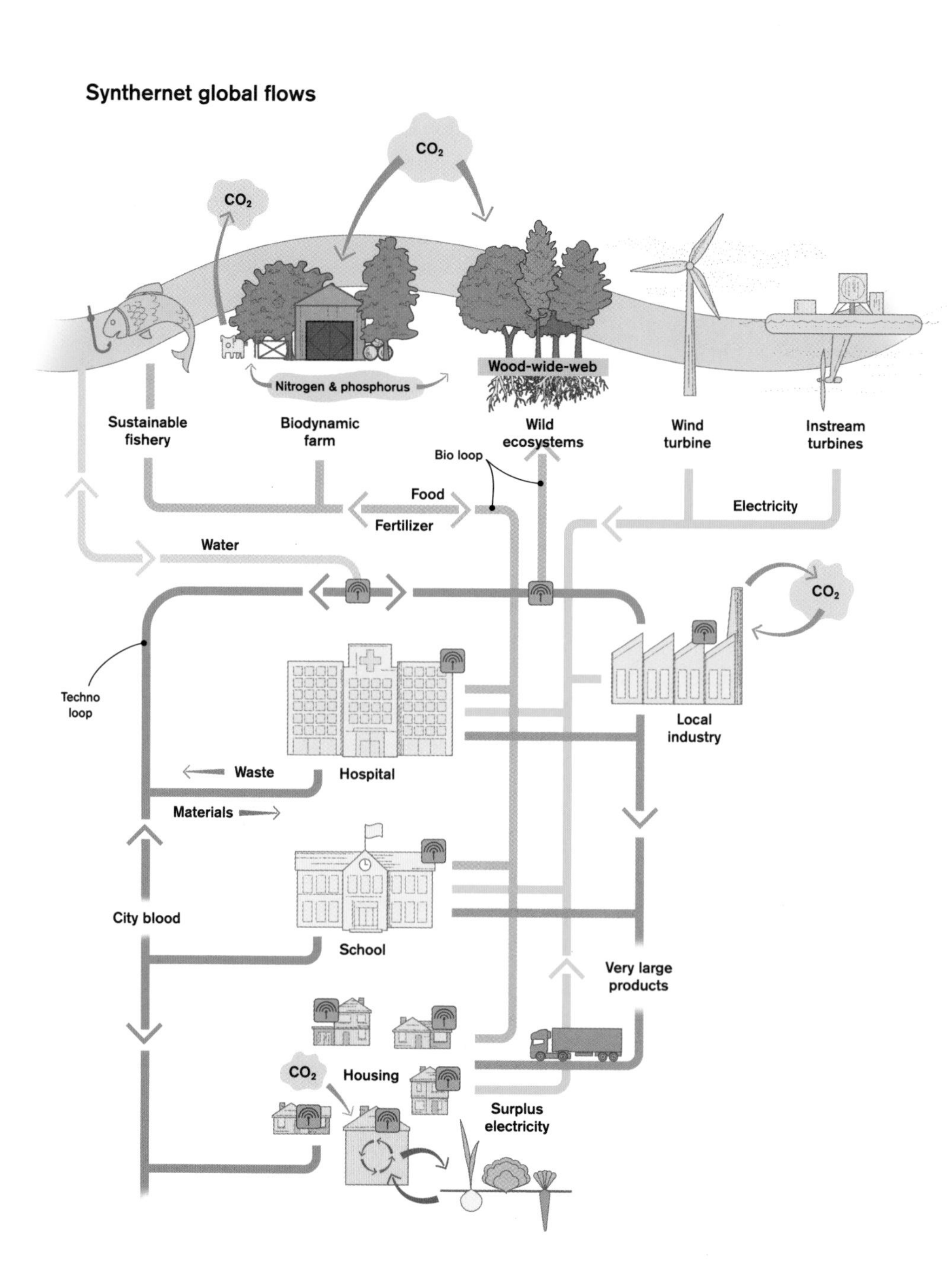

Peacefully coevolving with AI

Designing automated, self-regulating systems monitored and controlled by AI raises the question: How much control do we want to relinquish to artificially intelligent machines and how much oversight do we want to retain for ourselves? The more decision-making ability we give up, the more responsive and adaptive our new nomadic ecosystems could become, but it requires us to have a great deal of trust in the AIs we have developed.

Perhaps the simplest solution to this conundrum is to build machines that we like and trust. Maybe, if they were unique, and possessed emotional affection (or the robotic equivalent) for us, we could even learn to love them. We could create AIs that we viewed as our soulful, creative offspring, with spirituality and emotions. If we can have these capacities, then why can't they? Under such circumstances then, surely we would not hesitate in bequeathing our legacy to them, any more than we would to our children whom we love? Our approach to general AIs should be that of nurturing, caring parents, ensuring we give the AIs the examples and skills they need to make just decisions. And they must use this gift we give them to earn our trust and respect.

Our intelligent machines should need us as much as we need them—as mentors, guides, confidants, and hopefully friends. A beautiful example of such human/robot co-evolution is the creation of SHIMON, a jazz-playing narrow AI that can play the marimbas and improvise alongside a human performer. If a general AI could learn to do the same, then new friendships and alliances could be forged.

Opposite: We envisage that a synthernet technology stack installed in every building will manage the exchange between the biological and technical flows of the circular economy. Each building interacts with a stream of material that we call city blood—the technical loop—which is a supply of water and nutrients. Objects can be printed locally using materials extracted from city blood. Waste matter could be released to the city blood, where it circulates to places where such waste is useful. For example, large items such as cars, buses, even starships, might be manufactured at large local hubs. The buildings would be analogous to trees and the city blood to soil. Local microbial fuel cells could process domestic wastewater to create electricity, clean water, and organic waste, with any surplus exported to the grid or city blood. Biodynamic farms and fisheries supplement locally grown food. Organic waste is returned to the biological domain via synthernet routers. Wild ecosystems—managed at arms length—would provide leisure possibilities, and strictly limited food and materials supplies.

Above, top: An architecture and space research collective (Clouds Architecture Office/SEArch+) won the 3D Printed Habitat Challenge, organized by NASA and America Makes in 2019, with their technique for 3D printing with water ice, which could be used to construct short-term structures on cold moons or planets in the outer solar system.

Above: Tessellated Electromagnetic Space Structures for the Exploration of Reconfigurable, Adaptive Environments (TESSERAE) are modular, reconfigurable self-assembling buildings that could be used as in-orbit space modules or ground infrastructure. Their tiled components are reminiscent of the tesserae mosaics of ancient Rome.

Peaceful and cooperative coevolution of humans and AI is a pathway we must create together. We cannot expect to be unchanged by the arrival of machine intelligence; like becoming a parent, our lives will be altered in every possible way (if and) when it happens. We know we need to develop a circular economy and we need to build AIs to do it—we have no choice, but we must be wise about how we go about it. In the words of Bertrand Russell: *"Love is wise."*

A changing job market

One fear about increased automation is the emergence of extensive job losses. Decentralized manufacture within the synthernet will eliminate the need for lengthy and complex supply chains, simultaneously eliminating jobs in construction, mining, logistics, and distribution.

Previous waves of innovation have resulted in cultural shifts that tended to reduce menial and labor-intensive work in favor of more skilled and intellectual professions. Will the synthernet free even more of us to become artists, designers, musicians, or philosophers? Recall the words of science writer Arthur C. Clarke in 1969: *"The goal of the future is full unemployment, so we can play."*

One solution to many of the social and economic implications of our increasingly automated, digital world is that governments supply cash to every citizen sufficient to meet their basic needs—food, housing, utilities—using funds raised through a variety of mechanisms such as taxation or dividends from collective ownership of assets. Universal Basic Income (UBI) could bring about economic and social balance. Lifting people out of poverty would enable us to increase the number of asset-holders, so that society was comprised of millions of millionaires, rather than thousands of billionaires and billions of poor people. The ability to develop talents, rather than how well-off your parents were, would become the dominant factor determining a person's success.

Circular economics

Free markets are an essential part of the circular economy. Like foraging ants, a free market allows humans to explore new and interesting ways of adding value and trading, without unnecessary barriers and restraints. In chapter 9 we saw that excess materials could be traded within a local synthernet community, but they could also be sold to corporations and governments working on mega-projects such as orbital habitats or space colonies. And, in the same way that people have taken to developing algorithms

Universal basic income

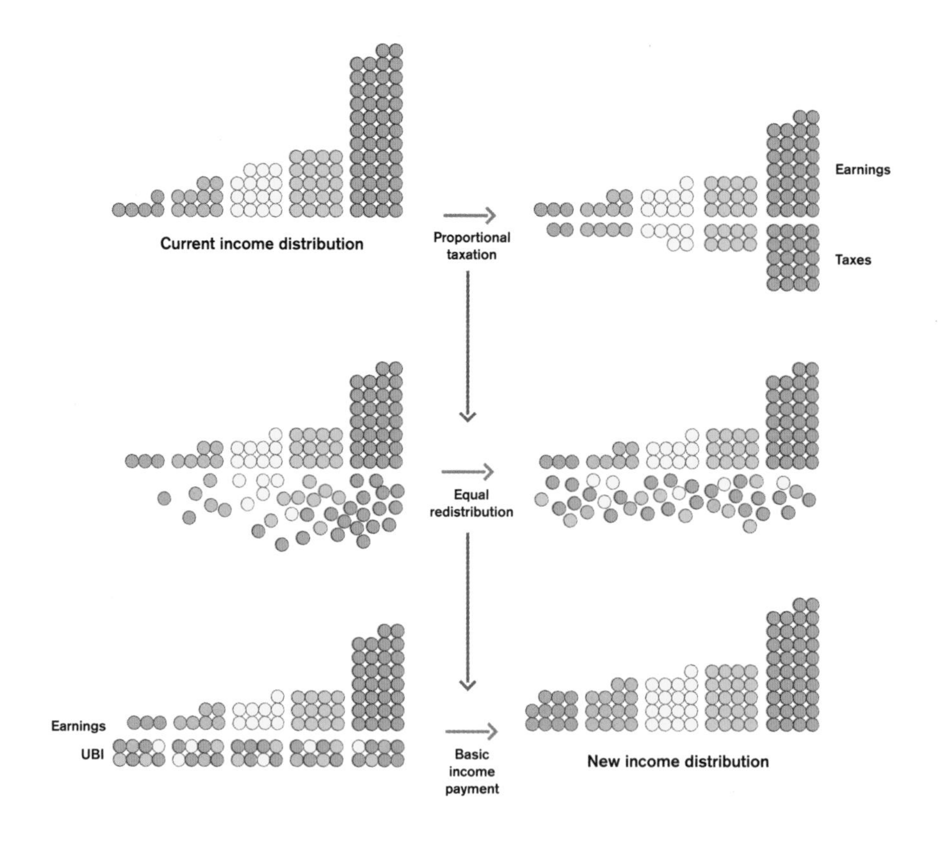

Above: UBI is a form of proportional taxation that is redistributed uniformly as tax credits, treating everyone identically while reducing income inequality. The benefit to higher earners of this additional taxation is the boost to the economy caused by eradication of poverty and a reduction in associated issues such as crime, mental illness, and poor health. UBI ensures everyone's basic income is covered so that, rather than fighting to survive, we can all spend time developing creative and professional skills or performing unpaid tasks such as parenting or caring for the elderly. UBI creates a safety net that allows everyone to take bigger risks and incentivizes employers to treat employees better, helping to ensure more equitable income distribution in the first place.

for apps on the internet, suddenly there would be a colossal space of design capability, open to everyone. Instead of relying on someone else to make the perfect thing that you need, you could make it yourself. Some people will be amazingly good at this, and their unique works will command great value. Of course, not everyone may have the desire, or talent, to develop truly innovative new designs, and individuals will be able to choose their level of involvement in the process—from passive customers of pre-tested designs using established synthernet materials from well-known brands, to active architects of not only the 3D-product design itself but the synthetic life forms that fabricate the AM feedstocks that will be used to build it.

Information ownership and decision-making in the digital age

The development of the internet has raised serious issues with existing notions (and laws) of intellectual property and copyright. Debates over the legality of, say, sharing the template for a 3D-printed weapon, scratch the surface of the new issues we will need to consider and possibly regulate or legislate in the synthernet.

In the digital age, an alternative mindset to that of intellectual property, patents, and copyright has emerged—the open-source movement. This refers to data, designs, products, software, and so on that is licensed for release into the public domain for free use and modification by anyone. In many cases, open-source software solutions have ousted traditional programs because the contribution of many independent developers across the world has made their code less prone to bugs and their development strategy more responsive to user needs. Similarly, large-scale public domain libraries such as Thingiverse and GitHub are helping level the playing field in 3D design.

With widespread roll-out of the synthernet technology, real-time connectivity, and millions of potential designers around the world, it will soon become impossible to monitor, let alone police design theft and counterfeiting. Instead we will have to reconsider our very relationship with information—we may need to accept a more open information economy, where value in a given product is generated by the work of many contributors, and efficiencies are found in rapid sharing of information in a collaborative society. Companies that once focused on developing niche proprietary designs may find an alternative business model in selling high-quality, extensively tested design templates that users can tweak at home.

Different approaches to democracy

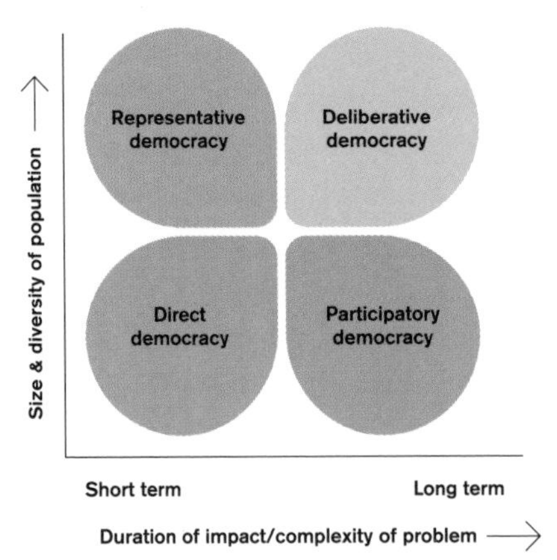

Left: Different kinds of democracy are appropriate for different situations. Our representative democracy works well for fairly complex short-term issues, such as deciding to go to war or setting tax budgets for the year. Climate change is long-term and complex in nature, and requires a different kind of democracy to solve it. Deliberative democracy selects representative groups of everyday people via democratic lotteries and gives them the resources—time and information—to make judgements on specific topics without interference from party politics and lobbyists. Other kinds of democracy such as direct democracy—in which everyone contributes—and participatory democracy—in which all affected members participate—are better suited to decisions that impact fewer people.

Big data and privacy

The rise of the internet and social media have also heralded a new era in citizens' concerns over privacy. Controversies like the British consulting firm Cambridge Analytica using Facebook data to help craft voting messages highlights the rising tensions between our desire to use big data to make our systems more intelligent and interconnected, and our need to shield ourselves from nefarious characters who want to mine that data and exploit it for personal advantage. Ultimately, these two aims are not mutually exclusive—it is merely an issue of information management—and as soon as we decide exactly what balance we, as a society, want to strike between big data science and privacy, technology will be able to provide the tools to implement it.

As we switch from large-scale centralized production to decentralized flows of materials, creating community wealth—for example, democratized business ownership and local investment in public resources—will be important for developing and maintaining social cohesion.

Governments may also find themselves in possession of new data and technology that can increase their control over populations, such as advanced AI surveillance systems or control over digital infrastructure. It is crucial that decision-makers regulate proactively, putting legislation and best practices in place early, to ensure society's most vulnerable are not disadvantaged as the synthernet starts to take shape.

A new politics for the digital age

AI is being used—for better or worse—to understand the needs, preferences, and vulnerabilities of individual voters and to influence their decisions. Benevolent use of this technology could encourage greater voter participation, but it is not hard to imagine a malevolent version of the same scenario, where selective voter suppression and inflammatory disinformation are used to favor a particular political party or influence a key referendum.

Maximizing democratic participation will help ensure that the transition to a circular economy is a force for equality and equity, and the internet and AI will be valuable assets in modernizing our democratic systems to involve more people. AI could offer near-instant fact-checking to limit the spread of disinformation and it could be employed to spot fraud and corruption, improving security and trust in a fully digital voting system that would boost the ease of participation.

In small groups—such as those of our nomadic, hunter-gatherer ancestors—problems can be addressed effectively with direct or participatory democracy. As group size increases, these approaches become cumbersome, so most modern democratic societies elect representatives. But representative democracy tends to be poor at resolving large-scale, long-term, and highly nuanced issues—how to implement a global synthernet economy, for instance—which might be best addressed with more participatory, deliberative decision-making processes.

Deliberative democracy emphasizes everyone's right and capacity to participate in and influence decisions, and the field has spawned the practice of "minipublics," in which groups of between 20 and 2,000 people come together to discuss a focus issue and consider solutions. These are intended to be more flexible than traditional politics, allowing space for different political viewpoints to contribute through different media—storytelling, humor, debates, even ceremonial speech—ideals that could be expanded using digital communication technologies to create huge online deliberative forums. The emphasis is on listening and finding common ground, rather than defeating an opposing point of view.

Above: The popular puzzle toy Rubik's Cube™ is a great analogy for the challenges facing humanity. Like the toy's colored faces, the number and type of atoms on Earth doesn't change (much) over time, and we need energy to rearrange them into different molecules. To solve the puzzle, you are supposed to move the faces until each side is a single color, but that is just one possible configuration of trillions that we could aim for. One of the frustrations on the way to solving a Rubik's Cube™ is that to reach the desired configuration we sometimes have to disrupt one part that we have already made; the same may be true for reaching a circular economy.

The challenge becomes how to manage conflicting opinions and ideas, and compile them into a final decision. By observing and learning from human examples of reaching consensus through well-moderated conversations, AI could be our strongest ally in this endeavor, if we can bring ourselves to trust it with such important work. For example, AI-powered tools such as Pol.is are platforms for managing and curating mass conversations. AI could also take greater responsibility for, say, automating simple local policy decisions or expert legal systems, supporting magistrates in courts, even triggering elections automatically based on petitions, protests, riots, public confidence or sentiment. Exactly how much control we want to delegate will be up to us.

Politicians need to be empowered to make long-term strategic policies, and form pan-partisan coalitions to maintain continuity of vision. At the same time, day-to-day policy-making needs to accelerate to keep pace with technological progress. Current decision-making systems were designed during the second industrial revolution at the end of the 19th century and rely on politicians having time to study an issue and develop a regulatory framework or policy change in response. In a rapidly evolving circular economy governments will need to become more agile and responsive to new developments and AI will be an essential ally in making this transition. We need to make sure that these agile governments are still beholden to the people, and the best way is to include everyday people in decision-making processes, for example, we could replace out-of-touch career politicians with everyday folk who are experts in the issues facing their own communities, chosen by civic lottery.

Making our way to a brave green world

Of course, there is more than one way that the technologies presented throughout this book or others could be used to build our future economies. We have only limited resources to fund scientific research, so how should we decide what will be done today? The direction that research takes—the building blocks we choose to invest in developing now—will determine our ultimate trajectory: toward a thriving circular economy, pre-industrial-era darkness, or a pillaged planet that offers nothing to sustain us. To make these decisions we must develop a shared vision of what kind of technology and society we want to build. We must collectively and individually decide what matters to "us."

In essence, this book is a quiet call to action. Everyone has a voice and can help to choose the direction that science should take and can consider how the decisions that they make can collectively impact the future. Technology is not an external force, driving our societies toward a predetermined future. Much of the fear of new technologies like robotics, genetic engineering, and AI is sparked by a feeling of powerlessness—that if we submit to them, we will become helpless riders on a runaway horse. But this outcome is only an inevitability if we fail to take the reins.

We are all capable of and responsible for guiding technological development through the decisions we make every day as citizens, consumers, workers, and investors. The biggest challenge we face as a species now is to reach a consensus on the values and principles on which we want our future societies to be built, so that we can shape the fourth industrial revolution to meet our future needs and dreams. Embracing novel, digital deliberation processes, and a highly participatory democratic system could be key to healing divisions and reaching such a consensus. It also requires that current decision-makers are willing and able to agree longer-term policy plans with their political opponents, backed by public consensus and supported by continued monitoring, which will facilitate the technological, economic, and societal changes that will pave the way for a circular economy.

What can I do?

As consumers, we can choose brands that boast transparent, sustainable supply chains. We can choose to bank, save, and invest our money in companies with higher ethical and environmental ratings, transparent supply chains, and commitments to waste-recapture or recycling schemes. We can support international, national, and local charities that run initiatives to decentralize manufacturing, teach repair skills, and encourage re-use of materials. As employees, we can pressure our senior colleagues to improve workplace circularity—remember that sustainability brings the double benefits of lower costs and a better reputation with environmentally conscious consumers. Small business owners can choose suppliers using renewable energy sources, sustainable manufacturing processes, or innovative waste-recycling schemes. The CEOs of large corporations have the opportunity to lead the way.

As voters, we can elect representatives who will fund circular economy programs, partner with industry to invest in sustainable programs, and create or support incentives for industry and academia to make data openly available. As constituents, we can encourage our elected officials to take steps to use more open deliberative democratic procedures, prioritize circular initiatives for city planning and infrastructure development, and build long-term pan-partisan coalitions locally and nationally, so that whoever is in power, we do not undo progress toward our shared long-term goals. We can urge our local representatives to vote for increased funding to support the development of new smart materials, new additive manufacturing processes, and new decentralized distribution models.

Like it or not we are all locked into a shared destiny—the destiny of planet Earth. If we are to reach a harmony with the resources and limits of planet Earth, we must work together and reach a shared vision for a future, where we each exist symbiotically with the biological world around us.

We are all empowered to decide what story our descendants will tell of us—let's make it a story that we all want to be a part of.

Above: Stefano Boeri Architetti's design for a "Smart Forest City" in Cancun, Mexico, which embeds a botanical garden within a high-tech city, sequestering CO_2 and cleaning the air as a result. These designs hint at what our *Brave Green World* might look like—a world where technology with biological levels of chemical and information storage and processing sophistication is integrated into the very fabric of the city.

Reality check

In chapters 2–9, we tried to illustrate how the ideas we have described throughout this book could be put into practice by talking about the production of an imagined future smartphone. We picked a smartphone—one of the most complex and ubiquitous objects that we currently make—on the assumption that if we can make a smartphone this way, we could likely build anything.

We discussed all the necessary subsystems of the biosmartphone and how those systems could be made and integrated in a novel growth-based manufacturing paradigm. We borrowed heavily from innovations discovered through biology and it culminated in the description of a manufacturing process involving droplet computers and synthetic cells. We went on to describe how such manufacturing would fit into a much larger global supply system and went through the process of ordering a device and requisitioning the necessary materials through the synthernet infrastructure. In doing so, we have explored some of the possibilities that are revealed by the unison of AM, synthetic biology, AI, and chemical computing. But how feasible are the ideas discussed and how long might we be waiting for such a biological smartphone to emerge?

Within the next decade, breakthroughs in flexible electronics and soft robotics may have made some of the first steps toward our biosmartphone. Cells are already being 3D printed into synthetic tissues; the top-of-the-range smartphones of 2030 might have touchscreens inspired by cephalopod skin. Their modular design would make their large components easy to remove and recycle, while artificially intelligent robots would disassemble the finer components.

By 2040, synthetic biology stacks, possibly based on fungi, could have become sophisticated enough to allow the manufacture of a wide range of commercial, consumer, and industrial chemicals, and the incorporation of a greater variety of crops engineered to produce materials and regenerate soil will increase revenues for farmers beyond just providing food. High-tech companies and researchers are already experimenting with new ways of performing additive manufacturing and growing devices and materials along the lines we have described. Such experiments, alongside ecosystem-design for space exploration over the coming decades, will filter into how we manufacture our goods with an ecosystem-based mindset.

By 2070, microgrid integration, combined with advances in ecosystem knowledge, additive manufacturing in harsh environments, and in-materio computation, could enable robust domesticated systems, like the heterotic computing stack that we have called the synthernet. It is reasonable to imagine that by this point science may have developed a fully integrated synthetic biology stack that can utilize chemical gradients to generate different phenotypes of genetically identical synthetic microbes, instructing them to build hierarchical materials and assemble them into 4D products. Whether a full smartphone could be produced this way is an open challenge.

Eventually all industries including agriculture, aerospace, water, energy, mining will be concerned with the atomic-resolution placement of material in their customers' environments and they will merge forces because they will find synergies at the molecular scale that will add great value to their companies. The smartphone of 2070 might therefore possess many of the traits that we have listed in the previous chapters and undoubtedly some we haven't thought of!

Above: The Daintree rainforest in Queensland, Australia demonstrates that the continuous recycling of materials between living organisms and the reservoirs of soil, air, and water around them can produce an intricate and beautiful ecosystem that can be sustained over a 100-million-year time scale. Just as birds inspired humans to fly, surely our industry can be inspired by the ecosystem-wide molecular technology that drives biological systems? Perhaps such bio-inspired technology can enable humanity to fulfil its potential, in harmony with natural systems.

References

Chapter 1

Page 12: K. E. Trenberth, J. T. Fasullo, J. Kiehl. *Bull. Am. Meteorol. Soc.* 2009 10.1175/2008BAMS2634.1

Page 12: R. Cassia *et al.* Front. *Plant Sci.* 2018 10.3389/fpls.2018.00273

Page 14: J. Rockström *et al. Nature* 2009 10.1038/461472a

Page 14: T. W. Lyons, C. T. Reinhard, N. J. Planavsky. *Nature* 2014 10.1038/nature13068

Page 14: Global Footprint Network 2019 http://data.footprintnetwork.org

Page 14: D. Moore *et al. Ecol. Indic.* 2012 10.1016/j.ecolind.2011.03.013

Page 15: N. S. Lewis, D. G. Nocera. *Proc. Natl. Acad. Sci. USA* 2006 10.1073/pnas.0603395103

Page 15: M. A. Green *et al. Prog. Photovoltaics Res. Appl.* (2020) 10.1002/pip.3228

Page 15: A. De Vos. *J. Phys. D. Appl. Phys.* 1980 10.1088/0022-3727/13/5/018

Page 22: R. L. Truby, J. A. Lewis. *Nature* 2016 10.1038/nature21003

Page 24: A. Yamaguchi *et al. Nucl. Eng. Technol.* 2017 10.1016/j.net.2017.02.001

Page 24: J. M. Pearce. *Int. J. Nucl. Governance, Econ. Ecol.* 2008 10.1504/ijngee.2008.017358

Chapter 2

Page 34: S. M. Sievert, C. Vetriani. *Oceanography* 2012 10.5670/oceanog.2012.21

Page 34: R. Mogul *et al. Astrobiology* 2018 10.1089/ast.2017.1814

Page 35: S. Yoshida *et al. Science* 2016 10.1126/science.aad6359

Page 35: S. Khan *et al. Environ. Pollut.* 2017 10.1016/j.envpol.2017.03.012

Page 43: C. Darwin *On the Origin of the Species* 1859

Page 43: D. Werth, R. Avissar. *J. Geophys. Res.* (2002) 10.1029/2001jd000717

Page 43: E. Salati *et al. Water Resour. Res.* 1979, 10.1029/WR015i005p01250

Page 46: S. C. Jung *et al. J. Chem. Ecol.* 2012 10.1007/s10886-012-0134-6

Page 48: H. Herz, W. Beyschlag, B. Hölldobler, *Biotropica* (2007) 10.1111/j.1744-7429.2007.00284.x

Page 50: C. Ratzke, J. Denk, J. Gore. *Nat. Ecol. Evol.* 2018 10.1038/s41559-018-0535-1

Page 50: C. O'Riordan, H. Sorensen 2008 10.1007/978-3-540-77949-0_12

Page 52: M. H. Braga *et al. J. Am. Chem. Soc.* 2018 10.1021/jacs.8b02322

Page 52: H. Sakai *et al. Energy Environ. Sci.* 2009 10.1039/B809841G

Page 52: J. C. Biffinger *et al. Biosens. Bioelectron.* 2008 10.1016/j.bios.2007.08.021

Chapter 3

Page 59: Chatham House (2020) http://resourcetrade.earth/

Page 62: T. A. Su *et al. Nat. Rev. Mater* 2016 10.1038/natrevmats.2016.2

Page 64: Y. Ding, D. Harvey, N.-H. L. Wang, *Green Chem.* 2020 10.1039/d0gc00495b

Page 67: World Resources Institute 2019 http://cait.wri.org

Page 68: V. Forti *et al.* 2020 http://ewastemonitor.info/

Page 70: Global Monitoring Laboratory www.esrl.noaa.gov/gmd/ccgg/trends/

Page 70: IPCC 2014 www.ipcc.ch/report/ar5/wg3/

Page 78: B. King, Y. Hu, J. A. Long. *Palaeontology 2018 10.1111/pala.12346*

Page 78: G. Falkenberg *et al. PLoS One* 2010 10.1371/journal.pone.0009231

Page 79: O. Kučera, M. Cifra. *J. Biol. Phys.* 2016 10.1007/s10867-015-9392-1

Page 79: J. Grinnell, K. McComb. *Anim. Behav.* 2001 10.1006/anbe.2001.1735

Page 79: M. T. Morita. *Annu. Rev. Plant Biol.* 2010 10.1146/annurev.arplant.043008.092042

Chapter 4

Page 80: I. Prigogine. *Is future given?* 2003

Page 83: C. Hidalgo. *Why information grows* 2015

Page 83: K. Webster. *The circular economy* 2017

Page 94: J. F. V. Vincent *et al. J. R. Soc. Interface (2006)* 10.1098/rsif.2006.0127

Page 98: N. R. Sinatra *et al. Sci. Robot.* 2019 10.1126/scirobotics.aax5425

Page 98: J. L. England. *Nat. Nanotechnol.* 2015 10.1038/nnano.2015.250

Page 98: C. Jarzynski. *Phys. Rev.* 1997 10.1103/PhysRevLett.78.2690

Page 100: I. Prigogine. *Science* 1978 10.1126/science.201.4358.777

Page 102: X. Zhou, Y. Hou. J. Lin *AIP Adv.* 2015 10.1063/1.4916886

Page 102: Y. S. Chen, M. Y. Hong, G. S. Huang, *Nat. Nanotechnol* 2012 10.1038/nnano.2012.7

Page 102: G. Reguera *et al. Nature* 2005 10.1038/nature03661

Page 102: W. M. Jacobs, D. Frenkel. *Biophys. J.* 2017 10.1016/j.bpj.2016.10.043

Page 102: N. Noor *et al. Adv. Sci.* 2019 10.1002/advs.201900344

Page 103: G. Villar, A. D. Graham, H. Bayley. *Science* 2013 10.1126/science.1229495

Page 103: J. M. A. Carnall *Science* 2010 10.1126/science.1182767

Page 103: A. J. Maheshwari *et al. Phys. Rev. Fluids.* 2019 10.1103/PhysRevFluids.4.110506

Chapter 5

Page 105: R. H. Carlson. *Biology Is Technology* 2010

Page 111: D. G. Gibson *et al. Science* 2010 10.1126/science.1190719

Page 112: J. Livet *et al. Nature* 2007 10.1038/nature06293

Page 113: F. J. M. Mojica *et al.* J. *Mol. Evol.* 2005 10.1007/s00239-004-0046-3

Page 113: M. Jinek *et al. Science* 2012 10.1126/science.1225829

Page 115: Y. Zhang *et al. Proc. Natl. Acad. Sci.* 2017 10.1073/PNAS.1616443114

Page 116: A. C. Komor *et al. Nature* 2016 10.1038/nature17946

Page 117: M. B. Elowitz, S. Leibier. *Nature* 2000 10.1038/35002125

Page 117: R. Daniel *et al. Nature* 2013 10.1038/nature12148

Page 117: B. A. Geddes *et al. Nat. Commun.* 2019 10.1038/s41467-019-10882-x

Page 121: C. A. Hutchison *et al. Science* 2016 10.1126/science.aad6253

Page 122: S. Basu *et al. Nature* 2005 10.1038/nature03461

Page 122: Y. Chen *et al. Science* 2015 10.1126/science.aaa3794

Page 126: S. L. Shipman *et al. Nature 2017 10.1038/nature23017*

Page 126: K. Chen *et al. bioRxiv* 2019 10.1101/857748

Page 127: C. E. Arcadia *et al. Nat. Commun.* 2020 10.1038/s41467-020-14455-1

Chapter 6

Page 128: Aristotle. *Metaphysics* 1908

Page 132: C. A. Solari *et al. Proc. Natl. Acad. Sci. USA* 2006 10.1073/pnas.0503810103

Page 134: L. Wolpert. *J. Theor. Biol.* 1969 10.1016/S0022-5193(69)80016-0

Page 135: M. B. Short *et al. Proc. Natl. Acad. Sci. USA* 2006 10.1073/pnas.0600566103

Page 136: R. K. Grosberg, R. R. Strathmann. *Annu. Rev. Ecol. Evol. Syst.* 2007 10.1146/annurev.ecolsys.36.102403.114735

Page 137: E. Davies *et al. PNAS* 2014 10.1073/pnas.1315080111

Page 138: S. Johnson. *Emergence* 2002

Page 140: K. Gödel. *Monatshefte für Math. und Phys.* 1931 10.1007/BF01700692

Page 140: A. M. Turing. *Proc. London Math. Soc.* 1936 10.1112/plms/s2-42.1.230

Page 151: R. Hovden *et al. Nat. Commun.* 2015 10.1038/ncomms10097

Page 151: S. H. Jeong *et al. APL Mater.* 2018 10.1063/1.4985754

Page 151: D. G. DeMartini, D. V. Krogstad, D. E. Morse. *Proc. Natl. Acad. Sci. USA* 2013 10.1073/pnas.1217260110

Chapter 7

Page 160: V. C. Müller, N. Bostrom. 2016 10.1007/978-3-319-26485-1_33

Page 161: Z. Liu *et al. Nano Lett.* 2018 10.1021/acs.nanolett.8b03171

Page 161: Q. Zhang *et al. Adv. Theory Simulations* 2019, 10.1002/adts.201800132

Page 161: T. Qiu *et al. Adv. Sci.* 2019 10.1002/advs.201900128

Page 161: M. A. Bessa, P. Glowacki, M. Houlder. *Adv. Mater.* 2019 10.1002/adma.201904845

Page 163: P. Rendell. *Turing Machine Universality of the Game of Life* 2015

Page 163: I. D. Couzin, N. R. Franks. *Proc. R. Soc. B Biol. Sci.* 2003 10.1098/rspb.2002.2210

Page 164: F. Delsuc. *PLoS Biol.* 2003 10.1371/journal.pbio.0000037

Page 169: L. Feijs, M. Toeters. *Int. J. Des. 12, 127-144* 2018

Page 170: J. Collins *et al. GECCO* 2018 10.1145/3205455.3205541

Page 174: A. E. Eiben, J. Smith. *Nature* 2015 10.1038/nature14544

Chapter 8

Page 181: E. Shapiro. *Interface Focus* 2012 10.1098/rsfs.2011.0118

Page 183: C. Horsman *et al. Proc. R. Soc. A Math. Phys. Eng. Sci* 2014 10.1098/rspa.2014.0182

Page 186: J. Von Neumann 1945 10.1109/85.238389

Page 189: A. Thompson. *Lect. Notes Comput. Sci.* 1997 10.1007/3-540-63173-9_61

Page 193: C. H. Bennett. *Int. J. Theor. Phys.* 1982 10.1007/BF02084158

Page 195: G. E. Crooks. *Phys. Rev. E.* 2 000 10.1103/PhysRevE.61.2361

Page 195: S. Y. Guo *et al. ChemRxiv* 2019 10.26434/chemrxiv.10250897

Page 198: A. C. Cavell *et al. Chem. Sci.* 2020 10.1039/c9sc05559b

Page 199: V. Kendon, A. Sebald, S. Stepney. *Philos. Trans. R. Soc. A Math. Phys. Eng. Sci.* 2015 10.1098/rsta.2014.0225

Chapter 9

Page 207: A. Alsbaiee *et al. Nature* 2016 10.1038/nature16185

Page 207: V. Meyer *et al. Fungal Biol. Biotechnol.* 2020 10.1186/s40694-020-00095-z

Page 210: Verisk Maplecroft *Waste Generation and Recycling Indices* (2019)

Page 212: Z. Liu *et al. Nat. Commun.* 2013 10.1038/ncomms2891

Page 219: A. Adamatzky, J. Jones. *Int. J. Bifurc. Chaos* 2010 10.1142/S0218127410027568.

Page 221: Valencia et al., *J.AGEE, 2016*, 10.1016/j.agee.2015.12.004

Page 223: Circle Economy 2020 www.circularity-gap.world

Page 223: D. Wiedenhofer *et al. Ecol. Econ.* 2019 10.1016/j.ecolecon.2018.09.010

Chapter 10

Page 226: K. Schwab. *The fourth industrial revolution* 2017

Page 231: J. F. Bastin *et al. Science* 2019 10.1126/science.aax0848

Page 235: R. Savery, G. Weinberg. In *Computer Simulation of Musical Creativity* 5 2018

Page 243: J. S. Fishkin. *Democracy and Deliberation* 1993

Page 243: B. Hennig. *The End of Politicians* 2017

Page 248: C. Yu *et al. Proc. Natl. Acad. Sci. USA* 2014 10.1073/pnas.1410494111

Page 248: F. Volpetti, E. Petrova, S. J. Maerkl. *ACS Synth. Biol.* 2017 10.1021/acssynbio.7b00088

Page 248: F. Y. X. Scott. *Virginia Polytechnic Institute and State University* 2017

Glossary

Adenosine triphosphate (ATP) An energy-carrying compound made of a ribose sugar attached to one adenine and three phosphate molecules.

Amino acid The building blocks of proteins, containing an amine, a carboxyl, and a unique side chain.

Biological loop One of two flows of matter in the circular economy that can be indefinitely regenerated using biological organisms.

Bottom-up process Higher order structure in a system that emerges spontaneously from interactions among the component parts.

Brownian motion The random movement of a particle in a fluid caused by bombardment from neighboring molecules.

Computable number Any real number that can be calculated to arbitrary precision by a finite terminating algorithm.

CRISPR-Cas9 A protein complex that can target and cut DNA sequences and insert them into the genome.

Deoxyribonucleic acid (DNA) The information storage molecule of cells—a double-helix of deoxyribose sugars and pairs of nucleic acids.

Dissipative system A thermodynamically open system, far from equilibrium, which exchanges material and/or energy with the environment.

Earthshine A colloquial phrase for Earth's thermal radiation.

Embedding The process of formulating a problem in a mathematical abstraction.

Encoding The process of writing a mathematically formulated problem into a physical medium.

Energy barrier The energy required to change the shape of a material.

Energy landscape A mathematical construct representing the potential energy of every possible configuration of a molecule.

Entropy The number of possible ways to arrange the components of a system.

Entropy budget The total change in entropy arising from the conversion of incident solar radiation to thermal radiation.

Enzyme A protein that binds to target molecules and facilitates a particular chemical reaction.

Eukaryote A biological kingdom including plants, animals, fungi, and yeasts.

Feedback loop When the outputs of a system are routed back as inputs, creating negative loops if the output suppresses its own production or positive loops if the output promotes production.

Feedstock The input material for a manufacturing or industrial process.

Generative design The process of applying an evolutionary principle to product design.

Genetic circuit A network of interconnected genes that dynamically sense and respond to the environment.

Genetic fitness The number of offspring that survive to adulthood produced by an organism with a particular genetic make-up.

Genotype The set of alleles (gene variants) possessed by a particular organism.

Geometry code (G-code) A standardized set of instructions typically used for CNC and 3D printing machines.

Gödel Number The numeral arising from the conversion of a logical proposition into a sequence of integers.

Heterotic computing A form of computing that blends different classes of computer into a single system.

Hierarchical material A material consisting of structural elements that themselves have structure, and so on.

In-materio computation The idea that formal computation can occur through combinations of physical processes inside a range of materials.

Messenger RNA (mRNA) Single-stranded RNA molecule transcribed from the genome, to convey a signal into the cell.

Metamaterial A highly structured material whose geometry creates novel properties.

Microgrid A local energy distribution system that can operate independently from the national grid.

Mitochondria Mitochondria are cellular organelles found in eukaryotic cells that provide energy for cellular processes.

Narrow AI A machine that performs specific tasks at human or super-human level, but cannot apply that intelligence to other scenarios.

Node An intersection in a network, such as a species in a food web or a single processing unit in an ANN.

Nomadic ecosystem A mobile set of closed material processes that can persist wherever they are located, if provided with sufficient energy.

Non-equilibrium physics The study of thermodynamic systems in which there is a flow of materials or energy with the environment.

Nuclear fission A reaction in which the nucleus of an atom splits into two (or more) lighter nuclei, releasing photons.

Nuclear fusion A reaction in which two (or more) atomic nuclei are fused. Nuclear fusion can be used to release energy.

Nucleotide (Base Pair) The structural components of DNA, made up of a base, a deoxyribose sugar, and a phosphate group. DNA is made up of Adenine, Cytosine, Guanine, and Thymine.

Oxidation A chemical reaction in which electrons are lost from an atom.

Phenotype The physical or behavioral manifestation of an organism's genotype, e.g. fur color or male courtship behaviour.

Pheromone A chemical secreted by an organism to send a message to another individual.

Photolithography The process of using light to impose a top-down pattern on a hard substrate.

Polymer Molecular chains consisting of repeating units. For example, a protein is a polymer of amino acids.

Polymerization The process of chemically joining molecules into repeating chains—polymers.

Prokaryote Living cells that lack internal membrane-bound organelles, such as bacteria.

Protease An enzyme that breaks down protein molecules.

Random access memory (RAM) A form of addressable computer memory that can be read and written in any order.

Reduction The process of adding an electron to a chemical species during a reaction.

Resistance A property of a conductor that determines the rate of flow of electrical charge per unit of electric driving force.

Respiration The metabolic process that oxidizes food to produce energy in the form of ATP.

Ribonucleic acid (RNA) A single-stranded polymer of nucleotides. RNAs play various roles in the cell.

Ribosome An enzyme that reads the nucleotide sequence of an mRNA strand and builds the corresponding amino acid sequence.

Soft matter Materials in which a small input of energy can result in a large change in shape or structure.

Steady state A form of non-equilibria where a flow of material or energy is constantly added to a system, which then leaves again.

Symbiosis An association between two organisms; may be mutually beneficial (mutualism), or harmful to one party (parasitism).

Synthernet A heterotic computing technology stack; the physical layer may include chemical or physical processes as part of the computational infrastructure.

Technical loop One of two flows of matter in the circular economy that consists of materials that are not biological in origin.

Thermodynamics (1st law) The principle in physics that energy cannot be created or destroyed.

Thermodynamics (2nd law) The principle in physics that entropy tends to increase in complex systems.

Top-down process A global process that is imposed on all the constituent parts at once.

Trabecular bone Highly porous bone tissue that retains the strength of solids with less material.

Transcription The first stage of gene expression, where a gene's DNA is copied into messenger RNA.

Transcription factor A protein that turns on or off the expression of a particular gene or group of genes.

Transfer RNA (tRNA) A ribonucleic acid molecule that physically links mRNA to the amino acid sequence of proteins.

Translation The second stage of gene expression, where messenger RNA is turned into a protein by the ribosome.

Trophic level Functional layers of the ecosystem network, such as primary producer, herbivore, and carnivore.

Turing complete A machine is Turing complete if it can emulate a Turing machine.

Wafer A slice of single crystal silicon on which microchips are fabricated.

Wood wide web The complex network of fungal hyphae that connect trees and plants beneath the ground.

Units of Measurement

Micrometer *μm* One thousandth of a millimeter (0.001mm)

Nanometer *nm* One millionth of a millimeter (0.000001mm)

Tons short tons, 2000 lbs

Tonne *t* metric tonne. 1000 Kg

Megatonne *Mt* 1 million metric tonnes

Gigatonne *Gt* 1 billion metric tonnes

Joule *J* A unit of energy

Calorie *ca* A unit of energy.

Megajoule *MJ* 1,000,000 joules

Volt *V* A measure of electrical potential, expressed as Joules per Coulomb

Watt *W* A measure of energy transfer, defined as one joule per second

Megawatt *TW* A million (1,000,000,000) watts

Terawatt *TW* A trillion (1,000,000,000,000) watts

Terabyte *TB* A unit of digital information. A terabyte is 1 trillion (1,000,000,000,000) bytes (B)

Exabyte *EB* An exabyte is 1 million (1,000,000) Terabytes (TB)

Hertz *Hz* The frequency of oscillation of a wave such as light, sound or water

Index

Acknowledgments

No book of this complexity is created in a vacuum and the hard work of many people are represented within its pages. We are indebted to UniPress, and particularly Kate Shanahan and Nigel Browning, for the opportunity to present these ideas. The many components of this book have been skilfully crafted into something besides the sum of its parts by Paul Palmer-Edwards under the watchful, curatorial eye of Natalia Price-Cabrera. We are very grateful for Rob Brandt's tireless efforts producing the marvellous high-quality illustrations that have brought the book to life, many of which are highly technical. We are extremely grateful to Ken Webster, Dan Widmaier, John Cumbers, Nick Asher, Kate "Scary" Oliver, Emma Byrne, James Lloyd, Stuart Nattrass, Andrew Cavell, and Tom Baden who provided invaluable feedback on early drafts of the book. Further thanks go to Linnea Lemma, Zvonimir Dogic, Tamily Weissman, Abby Birbach, Charlotte Mykura, Aurora Nedelcu, Maria Alejandra Mora-Sanchez, Gusz Eiben, Thomas Splettstoesser, Ed Cox, and Jack Collins, for their input on the visual elements of *Brave Green World*, and in particular, Chris Diewald, Julian Vincent, Gavin Munro, and Matt Thompson deserve special mention for going beyond the call of duty. We would also like to thank Krste Pangovski, Edon Vitaku, Harry Epstein and all at Nexus for providing a sounding board to help us to develop our ideas, alongside countless colleagues, students, family, and friends. Special mention must go to Randy Goldsmith and the ChemPU crew, especially Si Yue Guo for explaining the calculations behind chapter 8; and to Rachel Armstrong, whose work on Living Architecture was an early inspiration behind this book. Claire would like to say a special thank you to Seirian Sumner, Mark Maslin, Steve Cross, and everyone in the Talent Factory for all their encouragement and guidance with her career in science communication, and to Dan Chambers for his tireless support and understanding. Chris would particularly like to thank Jonathan Pritchard, Paul Barker, Bill O'Neill, Marc DesMulliez, David Wales, and Nathan Gianneschi, without whose guidance and support this book would not exist. And to Jessica Imboden, for her patience and unwavering encouragement.

Finally, it is with great sadness that we note the passing of Prof. Dame Georgina Mace, who was an inspiration to a generation of ecologists and who offered opportunities, support, and guidance at pivotal points in Claire's career. We are honoured that she was able to take time, amongst her many other roles and responsibilities, to read a draft and write an endorsement for our book.

Picture credits

p11: Courtesy of NASA Images. p13: Courtesy of NASA/SDO (top); courtesy of NASA/Goddard/MODIS Rapid Response Team (bottom). p26: Courtesy of EUROfusion/© Euratom. p29: Courtesy of Poravute Siriphiroon/Shutterstock. p40: Courtesy of Schmidt Ocean Institute (top); courtesy of Elif Bayraktar/Shutterstock (bottom). p47: Courtesy of Kichigin/Shutterstock. p49: Courtesy of Aguiar Lopes/Shutterstock (top); courtesy of Claire Asher (center left); courtesy of Fernando Calmon/Shutterstock (center right); courtesy Ilton Rogerio de Souza/Shutterstock (bottom). p50: Courtesy of Claire Asher. p59: Illustration data provided by Chatham House (2020) as part of the ResourceTrade.Earth project (http://resourcetrade.earth). p61: Courtesy of Nik Keevil/Shutterstock (top); courtesy of il21/Shutterstock (center left); courtesy of G-Stock Studio/Shutterstock (center right); courtesy of kim chul hyun/ Shutterstock (bottom). p65: Courtesy of Ian Dagnall/Alamy Stock Photo. p66: Courtesy of DBURKE/Alamy Stock Photo. p72: Courtesy of EMF and adapted from original by Ellen MacArthur Foundation. p79: Courtesy of Dennis Kunkel Microscopy/Science Photo Library. p87: © Arup and Davidfotografie (top); courtesy of MX3D/Joris Laarman Lab. p91: Courtesy of Carbon, Inc. p97: Illustration based on work by Julian Vincent et al, courtesy of Royal Society Publishing. p99: Courtesy of Nicole Hone, Victoria University of Wellington, New Zealand. p101: Images by Linnea Lemma. Thanks to Zvonimir Dogic. p110: Courtesy of Martin Fowler/Shutterstock (top left); courtesy of Chamille White/Shutterstock (top right); courtesy of Martin Krzywinski/Science Photo Library (bottom). p112: Courtesy of Frans Lanting, Mint Images/Science Photo Lab (top left); courtesy of Varvara Voinarovska/ Shutterstock (top right); courtesy of Tamily Weissman (bottom left)/Image by Tamily Weissman. The Brainbow mouse was produced by Livet J, Weissman TA, Kang H, Draft RW, Lu J, Bennis RA, Sanes JR, Lichtman JW. Nature (2007) 450:56-62; courtesy of Makoto Iwafuji/Eurelios/Science Photo Library (bottom right). p121: Courtesy of Thomas Deerinck, NCMIR/Science Photo Library. p125: Courtesy of Bolt Threads. p133: Courtesy of Aurora M. Nedelcu. p137: Courtesy of Alex Hyde/Nature Picture Library/Alamy Stock Photo. p139: © Chris Diewald. p144: Courtesy of Maria Alejandra Mora-Sanchez/Photography by Maria Alejandra Mora-Sanchez (top left & bottom left)/Photography by Brian Vogel http://brianvogel.photography/ (top right & bottom right). p147: Courtesy of Jose Luis Calvo/Shutterstock. P162: Courtesy of Maridav/Shutterstock (top); courtesy of PHOTOCREO Michal Bednarek/ Shutterstock. p165: Courtesy of Chris Forman. P166: Mark Brandon/Shutterstock (top); Kate Babiy/Shutterstock (bottom left); Tamara Kulikova/Shutterstock (bottom right). p169: Courtesy of Christiane M. Herr (top); courtesy of Alyssa M. Adams under Creative Commons License (bottom left); courtesy of Marina Toeters/Photography by Robin van der Schaft/Styling by Maaike Staal (© Marina Toeters) (bottom right). p170: Courtesy of Jack Collins/ © Commonwealth Scientific and Industrial Research Organisation, 2019. p172: Courtesy of Dagham Cam and Michail Desyllas at Ai Build Limited. p174: Original material courtesy of A. E. Eiben/Vrije Universiteit Amsterdam. p177: Original material courtesy of TeselaGen Biotechnology Inc., San Francisco CA. p178 & 179: Electrode design by Abhishek Sharma. p187: Courtesy of Alyssa and Paul Boswell. p188: Courtesy of Alan Richards photographer. From the Shelby White and Leon Levy Archives Center, Institute for Advanced Study, Princeton, NJ, USA. p190: Courtesy of LademannMedia-ALP/Alamy Stock Photo. p191: TE0808-04 "UltraSOM+ MPSoC Module with Xilinx Zynq UltraScale+ ZU15EG," courtesy of Trenz Electronic GmbH, Hüllhorst, Germany, 2020. p198: Courtesy of Chris Forman. p201: Courtesy of Chris Forman. p202 & 203: Illustration created with permission from TEZ ARCHITECTS. p206: Courtesy of jonathan-filskov-photography/iStock. p214: Courtesy of Rachel Armstrong. p219: Courtesy of Andrew Adamatzky. p221: Courtesy of Volodymyr Kalyniuk/Alamy Stock Photo (left). p225: Courtesy of Gavin & Alice Munro/Full Grown Ltd/Photos by Chris Robinson. p232: Courtesy of U.S. Geological Survey, Department of the Interior/USGS (top); courtesy of NASA Images (bottom). p235: Courtesy of Juergen Held/travelstock44/Alamy Stock Photo. P238: Courtesy of Clouds Architecture Office/SEArch+ (top); courtesy of Ariel Ekblaw, MIT Media Lab (bottom). p242: Illustration interpreted from original material courtesy of Royal Society for the encouragement of Arts, Manufactures and Commerce. p244: Image by scistyle.com, courtesy of Chris Forman. p247: Cancun Smart Forest City, courtesy of Stefano Boeri Architetti and The Big Picture. p249: Courtesy of noexcuseG/Shutterstock.